Analytical Methods for Polymer Characterization

Analytical Methods for Polymer Characterization

Rui Yang

CRC Press

Taylor & Francis Group
Boca Raton London New York

CRC Press is an imprint of the
Taylor & Francis Group, an **informa** business

CRC Press
Taylor & Francis Group
6000 Broken Sound Parkway NW, Suite 300
Boca Raton, FL 33487-2742

First issued in paperback 2020

ISBN-13: 978-0-367-57235-8 (pbk)
ISBN-13: 978-1-4822-3354-4 (hbk)

Library of Congress Cataloging-in-Publication Data

Names: Yang, Rui (Chemical engineering professor), author.
Title: Analytical methods for polymer characterization / Rui Yang.
Description: Boca Raton : CRC Press, 2018. | Includes bibliographical references.
Identifiers: LCCN 2017039410 | ISBN 9781482233544 (hardback : alk. paper)
Subjects: LCSH: Polymers--Analysis--Problems, exercises, etc. | Chemistry,
Analytic--Problems, exercises, etc.
Classification: LCC QD139.P6 Y36 2018 | DDC 547/.7046--dc23
LC record available at https://lccn.loc.gov/2017039410

Contents

Preface

During the development of polymer materials in past decades, various existing characterization techniques were applied to investigate the structures and properties of polymer materials. At the same time, some new techniques were developed. To summarize these achievements, many books on principles, instruments, and applications of some specific characterization techniques were published. However, there are few textbooks for undergraduates that introduce the common analytical methods of polymer materials.

Therefore, based on eight years of teaching handouts, I finished this textbook, which contains chromatographic methods (gas chromatography, inverse gas chromatography and pyrolysis gas chromatography), mass spectrometry, spectroscopic methods (ultraviolet-visible spectroscopy, infrared spectroscopy, Raman spectroscopy and nuclear magnetic resonance), thermal analysis (differential scanning calorimetry and thermogravimetry), microscopy methods (light microscopy, scanning electron microscopy, transmission electron microscopy and atomic force microscopy), and X-ray diffraction. This textbook is for undergraduates who are majoring in polymer materials. From this book, they will learn the principles, instruments, sampling and applications of common methods that are generally used in the research of polymer materials.

Due to limits in my own knowledge and understanding, there might be mistakes and omissions in this textbook. Any constructive advice and comments are welcome and greatly appreciated.

I wish to dedicate this book to Prof. Kunhua Wang, my tutor. She always gave me suggestions and support when I had problems or was puzzled. I would like to thank my colleagues at Tsinghua University for the IR spectra and SEM, TEM, and AFM photos. Finally, I want to thank my parents, my husband, and my son. They are forever the driving force in my life.

Rui Yang
Tsinghua University
August 3, 2017

Author

Dr. Rui Yang earned her BS, MS and PhD in polymer materials at the Department of Chemical Engineering at Tsinghua University in 1994, 1996, and 2005, respectively. She began her teaching and research in the Department of Chemical Engineering at Tsinghua University in 1996 as an assistant professor, and became an associate professor of polymer materials in 2004. Her research interests include aging behavior, mechanisms, and stability evaluation of polymer materials and their lifetime prediction; the latent heat phase change material (PCM) composites for solar energy storage, thermal regulation, medical treatment, etc.; and the structure-property relationship of polymer materials. She has published over 100 peer-reviewed papers and was awarded the Second Prize for the Natural Science Award of Ministry of Education of China in 2005, and the Second Prize for the Science and Technology Progress Awards of Beijing in 1999. She is the board member of Materials Division in Chinese Mechanical Engineering Society (CMES), and the board member of Polymer Materials and Engineering Division in Chinese Materials Research Society (CMRS). In her research, she has accumulated experience in many analytical instruments and has become an expert in Fourier transform infrared spectroscopy (FTIR), pyrolysis gas chromatography (PGC), as well as their applications in polymer materials. She has been teaching instrumental analysis of polymer materials for undergraduates at Tsinghua University since 2010, and was awarded the Tsinghua University Young Teacher Teaching Excellence Award. She has participated in the publication of a Chinese textbook *Advanced Instrumental Analysis of Polymers* and has completed the revision of the third edition. Currently, this textbook has been used by more than 40 universities and institutes.

Introduction

The development of new materials reflects the technological and scientific status of a country. Among the various new materials, polymers form one of the most active categories, and have attracted increased attention because of their light weight, high strength-to-weight ratio, good impact endurance, good processibility, and, most importantly, great functional potential in future industries. Nowadays, based on the structure–property relationship, a new polymer material is being designed and prepared. Aiming at a desired property, advanced analytical instruments are necessary for its structure determination.

As opposed to metals, ceramics, and small molecular organics, polymer materials have specific features regarding their composition, structure, and properties. Accordingly, specific analytical instruments, sampling techniques, spectra interpretation, and data processing are required for polymer materials.

Questions in Polymer Materials Research

During polymer materials research, we often encounter the following questions:

- Is the structure of a new polymer the same as was anticipated? Are the average molecular weight and distribution satisfactory?
- Do its aging stability and its mechanical, rheological, thermal, electronic, and optical properties meet practical demands?
- What is the relationship between structure and performance? How is this relationship adjusted?
- How will quality control and product evaluation be carried out during the preparation and processing of the polymer material?
- What happens during its storage and application?

To answer these questions, the first step is to understand the research subject: polymer materials. Thereafter, one must learn to use analytical tools, including principles, sampling, and data interpretation. Only through the combination of understanding of polymer materials and analytical tools can we achieve our research target.

About Polymer Materials

There are various kinds of polymer materials. Generally, polymer materials can be classified as plastics, rubbers, fibers, coatings, and adhesives according to their application. They may also be classified as general polymers, high performance polymers, and functional polymers (such as biomimetic materials, biomedical materials, and optoelectronic polymers) according to their function.

A polymer material is a complex system. The polymer could be a homopolymer, copolymer, or blend. Furthermore, it may contain an emulsifier, molecular weight modifier, chain transfer agent, and chain terminator. Some other additives such as inorganic fillers, plasticizers, stabilizers, pigments, and antistatic agents may be introduced into the material; in addition, there could be a solvent and residual monomers.

Polymer materials are quite different from metals, ceramics, and small molecular organics. They have the following features:

- Polymer materials have a very high molecular weight, which is not a definite number, but an average of molecular weights of all macromolecules. Therefore, the physical properties of the polymer (such as melting temperature, glass transition temperature, and strength), are not definite and change with molecular weight.
- The chemical structure can be easily modified by changing the monomers or by reacting them with pendant groups. This will also change the polymer's properties. Consider a long molecular chain with hundreds to thousands of pendant groups: there is an almost infinite number of possible chemical structures.
- Polymer materials have elasticity as well as viscosity. Although this feature endows the polymer materials with strength and modulus, these properties change with time and temperature. One must pay attention to this feature during the practical application of polymers.
- The physical state of a polymer material changes with environmental conditions such as temperature, pressure, frequency, and media. Therefore, the physical properties change accordingly.
- Polymer materials exhibit composition and structural diversity. There are homopolymers, copolymers, and blends; filled composites and fiber reinforced composites; and multilayer membranes and multilayer coatings. Even in a pure polymer, there are crystal and amorphous domains. These multiphase and multicomponent features endow polymer materials with various functions.

The chemical composition, structure, processing, and properties of polymer materials have been introduced in *polymer chemistry, polymer physics*, and *polymer processing*. This book focuses on the characterization of polymer materials by analytical instruments.

Chain Structure of Polymer

The chain structure of a polymer refers to the chemical structure of a single macromolecular chain, which determines the basic properties of a polymer material.

■ Chemical Structure, Molecular Weight, and Distribution

Once the chemical structure of a polymer changes, its properties vary greatly. For example, polyethylene (PE) is a tough, crystallizable plastic. If one hydrogen atom in PE is substituted by a benzene ring, polystyrene (PS) is obtained, which is a brittle plastic with high strength, but it cannot be crystallized in general. If one hydrogen atom in PE is substituted by a carboxyl group, polyacrylic acid (PAA) is obtained, which is a water-soluble polymer, while PE and PS are insoluble in water.

The molecular weight of a polymer greatly influences its strength and processibility. A high molecular weight corresponds to high strength and weak processibility; a low molecular weight corresponds to low strength and good processibility. The molecular weight and distribution of a polymer must be adjusted according to its processing mode and application.

■ Bonding Mode

There are two bonding modes in mono-substituted monomers: the typical head-to-tail bonding and some head-to-head bonding. A polymer with the latter bonding has poor stability and is easily degraded from these points.

$$\text{wwww } CH_2\text{-}CH\text{-}CH_2\text{-}CH \text{ wwww} \qquad \qquad \text{wwww } CH_2\text{-}CH\text{-}CH\text{-}CH_2 \text{ wwww}$$
$$\qquad \quad | \qquad \quad | \qquad \qquad \qquad \qquad \qquad | \quad | $$
$$\qquad \quad R \qquad \quad R \qquad \qquad \qquad \qquad \qquad R \quad R $$

Head-to-Tail Bonding   Head-to-Head Bonding

■ Branching and Crosslinking

If monomers with more than two functional groups exist, or if chain transfer occurs during polymerization, branching structures are formed. A branched polymer behaves quite differently from a linear polymer. If the branching degree is further increased, a crosslinking polymer may be produced, which is neither soluble in a solvent, nor fusible when heated.

■ Stereoregularity

For a mono-substituted vinyl polymer, such as polypropylene (PP), there are three steric structures, i.e., isotactic, syndiotactic and atactic. Isotactic and syndiotactic PPs are plastics, while atactic PP is a viscous semisolid.

Isotactic: Substituents are distributed on the same side (Red–C atom; White–H atom; Blue–substituent group; the same below)

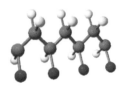

Syndiotactic: Substituents are distributed alternately on both sides

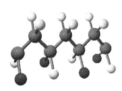

Atactic: Substituents are randomly distributed

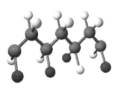

The configuration that the adjacent two substituents are on the same side is labelled as *m*; the configuration that the adjacent two substituents are not on the same side is labelled as *r*. So isotactic polymer is labelled as *mmmm*; syndiotactic polymer is labelled as *rrrr*.

■ Cis-trans Isomerism

If there are double bonds in a polymer, cis- or trans- isomers exist. For example, cis-polybutadiene (PB) is a rubber, while trans-PB is a plastic.

■ Sequence Distribution of Copolymer

According to the reactivity ratios and feeding sequence, the copolymerization of two or more monomers gives rise to copolymers with many sequence distributions, and thus, different properties.

Random Copolymer: –AABAAABBABBBAA–

Alternating Copolymer: –ABABABABABABAB–

Block Copolymer: –AAAAAAAABBBBBB–

Graft Copolymer:

```
—— AAAAAAAAAAAAAAAAAAAAA ——
        |              |
        B              B
        |              |
        B              B
        |              |
        B              B
        |              |
        B              B
        |              |
       ...            ...
```

Aggregation Structure of a Polymer

The aggregation structure of a polymer, which determines its practical performance, represents the manner in which the macromolecular chains are arranged together.

- Physical State of a Polymer

    The physical state of a polymer depends on the extent of movement of the macromolecules. When the temperature is elevated, a polymer material changes from a glassy state to a rubbery state; and, finally, to a viscous state. In the glassy state, almost no molecular movement is determined. In the rubbery state, the movement of segments predominates, although the macromolecule chains still cannot move. In the viscous state, all macromolecular chains can move.

- Crystallization

    Many polymers can crystallize by regular chain folding. However, other polymers cannot crystallize. Even in crystallized polymers, there are amorphous regions. Crystal phases endow polymer materials with a high strength, while amorphous phases endow polymer materials with impact endurance. The optimal combination of crystal and amorphous phases makes a polymer material strong and tough.

    Some polymers maintain a partially ordered crystal structure (anisotropy) even in melts or solutions. These polymers are liquid crystal polymers, with both crystal and liquid properties.

- Orientation

    Macromolecule chains align along external forces to partially ordered and anisotropic structures. This phenomenon is called orientation. The orientation of segments can be realized in the rubbery state. The orientation of macromolecules can be realized in the viscous state. An oriented structure is thermodynamically unstable and can only be maintained with external forces. Disorientation happens naturally once the external force is removed. Therefore, it is necessary to "freeze" oriented structures at low temperatures to maintain the orientation.

- Blends and Alloys

    Two or more polymers can blend into an alloy: generally, mixing at the molecular level cannot be realized. A polymer and inorganic filler can blend into a composite. Both alloys and composites are multiphase and multicomponent materials.

Reaction and Changes of Polymer Materials

Over the lifetime of a polymer material, various reactions and changes can occur during polymerization, processing, and application; these reactions and changes must be characterized in detail.

During polymerization, monomer(s) polymerize to a polymer. To achieve this, the raw materials are analyzed for purity; the synthesis conditions are optimized; the chemical structure of the products are characterized; the molecular weight and distribution are determined; and the branching/crosslinking structures are analyzed. For a copolymer, the composition and sequence distribution are determined.

During processing, a polymer is processed to a product. For this, the processing conditions are optimized; and the microscopic phase structure and morphology are observed. For reactive processing, chemical reactions such as grafting and degradation are characterized.

During the storage and application of a product, its properties deteriorate with time (aging). For some materials, high stability and a long lifetime are expected. For other materials, quick degradation after usage is expected for recycling or reducing pollution. Therefore, changes in the structure and properties of products with time are to be monitored, and the degradation mechanisms are to be investigated.

Structure–Property Relationship

For a polymer material, its properties are a concern. The basic properties of a polymer depend on its chain structure, while the practical properties depend on the aggregation structure. For various polymer materials, the properties are different. For a given polymer material, different practical properties may be obtained by different processing techniques. For a required property, various polymer materials may be used. The properties of polymer materials are determined not only by their chemical structure, but also by their physical structure (morphology). This is called a structure–property relationship.

Analytical instruments help researchers to investigate chain and aggregation structures, and their changes. Based on the structure information and practical properties of a polymer, a structure–property relationship can be established, which is quite important for the design, preparation, and application of polymer materials.

Analytical Methods of Polymer Materials

The chain structures of polymers focus on functional groups in single molecular chains. Therefore, analytical methods such as pyrolysis gas chromatography–mass spectrometry (PGC-MS), Fourier transform infrared spectroscopy (FTIR), ultraviolet–visible spectroscopy (UV–Vis), Raman spectroscopy, and nuclear magnetic resonance (NMR) are often used.

The aggregation structures of polymers depend on the relative arrangement of the molecular chains, reflected by the interactions between the functional groups. Therefore, analytical methods such as FTIR, differential scanning calorimetry (DSC), scanning electron microscopy (SEM), transmission electron microscopy (TEM), and x-ray diffraction (XRD) are often used.

The reaction and changes of polymers depend on changes in the functional groups, and changes in the interactions between functional groups accompanied with morphological changes. Therefore, all analytical methods for chain structure and aggregation structure characterization can be used.

Before the analysis of polymer materials, one of the first important tasks is to collect sample information, including the source of the sample, the areas in which it is used, its appearance, etc. According to these, some primary conclusions can be drawn: whether the sample is a thermoplastic or a thermoset, and if the sample is filled with inorganic matter. Sometimes, it is even possible to determine the exact polymer. Based on this information, proper analytical methods are chosen according to the purpose (e.g., the information that is to be derived about the sample).

The second important task is to obtain adequate knowledge of analytical methods. Analytical instruments are powerful tools, but only if a person clearly knows what every instrument can do, how it is operated, and what information it can supply. With this knowledge, a person can choose the proper instruments and sampling methods, and obtain as much information as possible.

The third important task is to integrate enough data and information in order to see the whole picture. In many cases, an analytical problem can be solved by various instrumental methods. It is important to carefully analyze the viewpoint of each method and what are the similarities and differences between them. In other cases, a complex problem needs integrating information from many methods. How to integrate this information together is a big problem.

In summary, the successful instrumental analysis of polymer materials requires a comprehensive analysis from various viewpoints, and on different levels. No single instrument is appropriate for all purposes. Every instrument has its advantages and disadvantages. It is the researcher's responsibility to choose the proper instrument and obtain satisfactory data at a reasonable cost. It is important to remember that an expensive instrument is not the optimal choice, nor is a popular or even an advanced instrument.

Chapter 1

Gas Chromatography and Inverse Gas Chromatography

1.1 Introduction

In 1906, M. S. Tswett, a Russian botanist, first developed the technique of chromatography while investigating the separation of plant pigments. In his experiment, a glass column was filled with calcium carbonate particles as the stationary phase, and a mixture of pigments was introduced from the top of the column. Next, petroleum ether, as the mobile phase, was added to the column. Gradually, pigments with various colors separated (Figure 1.1), and hence, "chromatography" was named. As chromatography, which has long been employed to separate various mixtures, has developed considerably, it is no longer limited to colored samples.

A chromatogram is a plot of signal intensity from a detector versus time; it shows the results obtained from chromatographic analysis. Figure 1.2 shows a sample chromatogram. The straight line OCD is the baseline, corresponding to the signal without samples. The peak CED corresponds to the presence of a component. EB, the distance from the peak top to the baseline, is the peak height (h). Tangents to two sides of the peak meet the baseline at I and K, and the distance between I and K is called the peak width (W). The width at $1/2\ h$ is called the half-peak width ($W_{1/2}$). When a sample is injected into the column, the time taken by it to pass through the column is called the retention time (t_R), which consists of two parts. The first part t_M is merely the time required for a component to pass from one end to the other end of the column as if it were not dissolved in the stationary phase. The second and the more important part t'_R is the time taken for the component

1

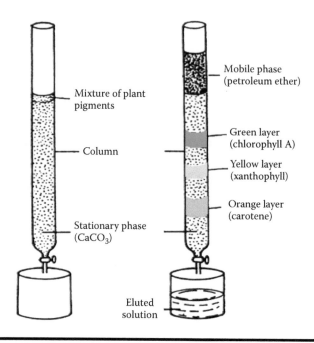

Figure 1.1 Chromatographic analysis of plant pigments by Tswett.

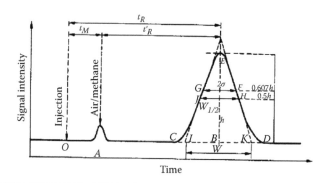

Figure 1.2 Sample chromatogram.

to move along the column while it is continuously dissolving in, and diffusing out, of the stationary phase.

As shown in Figure 1.1, components in a mixture can be separated through a column. How does this occur? Figure 1.3 shows the schematic of the separation. First, the mixture is at one end of the column. With the help of the mobile phase (typically, a carrier gas or solvent), the mixture moves along the column. As various components exhibit different interactions with the stationary phase, they exhibit

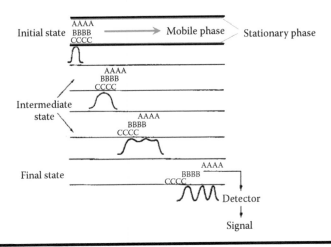

Figure 1.3 **Schematic of the separation of a mixture through a column.**

different partitions between the stationary and mobile phases. Hence, the partition coefficient K is defined as follows:

$$K = C_L / C_G. \tag{1.1}$$

Here, C_L and C_G represent the concentrations of a component in the stationary and mobile phases, respectively.

At a specific temperature in the equilibrium state, the partition coefficient (K) is decided by the thermodynamic properties of the component, as well as the stationary and mobile phases. It is possible to separate the two components as long as they exhibit different K values, even though the difference is extremely small. Therefore, from a theoretical viewpoint, it is always possible to find appropriate stationary and mobile phases to separate two similar components. On the other hand, if both the components exhibit the same K value, they cannot be separated. A component with a high K value moves slowly in a column.

In practice, the possibility of chromatographically separating two components also depends on the resolution R:

$$R = 2\left(\frac{t_{R2} - t_{R1}}{W_1 + W_2}\right). \tag{1.2}$$

Here, t_{R1} and t_{R2} are the retention times of two components, respectively, and W_1 and W_2 are their corresponding peak widths.

Figure 1.4 shows three typical cases. When R < 1, the two components overlap and cannot be separated (Figure 1.4a). When R ≥ 1.5, the components are

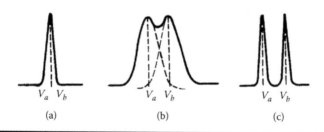

Figure 1.4   Illustrations of peaks with different resolutions. (a) R < 1, (b) 1 ≤ R < 1.5, and (c) R ≥ 1.5.

completely separated (Figure 1.4c). As shown in Figure 1.4b, partial separation is observed when 1 ≤ R < 1.5.

According to the different mobile phases, chromatography can be categorized into gas chromatography (GC; with carrier gas as the mobile phase) or liquid chromatography (LC, with solvent as the mobile phase). In this chapter, GC and inverse gas chromatography (IGC) with a carrier gas as the mobile phase are discussed.

1.2 Gas Chromatography

1.2.1 Theory

From Equation 1.2, if complete separation needs to be achieved, i.e., good resolution, the difference between the retention times of two components must be as large as possible. In addition, the peak widths must be as small as possible. Hence, a suitable separation column is required with a high separation efficiency, as well as optimal operating conditions. All of these problems can be solved according to the plate and speed theories. For clarity, only isothermal operating conditions are discussed.

1.2.1.1 Plate Theory

Let us consider a GC column as a distillation column with n plates, with H being the height of each plate. Separation can be considered as a continuous dissolution equilibrium between the stationary and mobile phases on every plate. Some assumptions have been reported [1]:

1. A linear partition isotherm exists between the two phases.
2. Partition equilibrium of a component between the stationary and mobile phases is achieved instantly.
3. Absence of longitudinal diffusion for the component in the gas phase.
4. Incompressible carrier gas.

Based on the above assumptions, the elution profile of the component can be expressed by a Gaussian function (Equation 1.3), which relates the component concentration C to time t:

$$C = \frac{\sqrt{n}C_0}{\sqrt{2\pi}t_R} \exp\left[-\frac{1}{2}n\left(1-\frac{t}{t_R}\right)^2\right].$$ (1.3)

Here, n is the plate number, and C_0 is the sample mass. The standard deviation σ in Equation 1.3 is expressed as follows:

$$\sigma = t_R/\sqrt{n}.$$ (1.4)

Hence, Equation 1.3 becomes

$$C = \frac{C_0}{\sigma\sqrt{2\pi}} \exp\left[-\frac{(t-t_R)^2}{2\sigma^2}\right].$$ (1.5)

When $t = t_R$ and $C = C_{max}$, the component concentration attains the maximum, i.e., the peak height h:

$$h = C_{max} = \frac{C_0}{\sigma\sqrt{2\pi}}.$$ (1.6)

The peak width can be expressed as follows:

$$W = 4\sigma.$$ (1.7)

The intensity of a chromatographic peak can be expressed by h and W. W represents the movement of the component, i.e., the diffusion in the gas and stationary phases, and it is affected by the operating conditions. W is an important parameter related to the separation efficiency.

Typically, the performance of a GC column is expressed by the theoretical plate number n (Equation 1.8) or the effective plate number n_{eff} (Equation 1.10) and the theoretical plate height H (Equation 1.9).

$$n = 5.54\left(\frac{t_R}{W_{h/2}}\right)^2 = 16\left(\frac{t_R}{W}\right)^2 = \left(\frac{t_R}{\sigma}\right)^2$$ (1.8)

$$H = L/n \tag{1.9}$$

$$n_{eff} = 5.54 \left(\frac{t'_R}{W_{h/2}} \right)^2 = 16 \left(\frac{t'_R}{W} \right)^2 = \left(\frac{t'_R}{\sigma} \right)^2. \tag{1.10}$$

Here, L is the length of the column. The higher the values of n and n_{eff} and the lower the value of H, the better the column performance.

From Equation 1.8, under isothermal conditions, W is proportional to t_R and is inversely proportional to n. With increasing analysis time, the W of a component gradually increases.

1.2.1.2 Speed Theory

In an ideal column, the peak shape is quite similar to the input distribution. Peak broadening is not observed during the elution of components. In actual columns, broadening occurs because of turbulence diffusion, longitudinal diffusion, and mass-transfer resistance. Peak broadening leads to decreased separation efficiency. Figure 1.5 shows the change of H with the flow rate of the carrier gas measured by experiments. J. J. Van Deemter has proposed Equation 1.11 to explain the relationship between the resolving power of a chromatographic column to the various flow and kinetic parameters:

$$H = A + B/\bar{u} + C\bar{u}. \tag{1.11}$$

Here, $\bar{u}$ is the average flow rate of the carrier gas in cm/s, which can be calculated by $\bar{u} = L/t_M$.

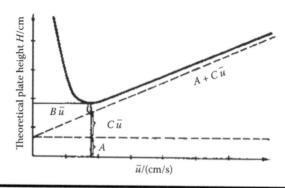

Figure 1.5 Relationship between the theoretical plate height and carrier gas flow rate.

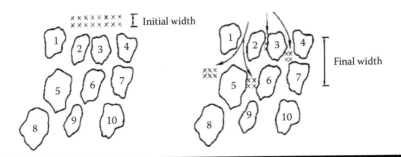

Figure 1.6   Schematic of turbulent diffusion

In Equation 1.11, A represents the peak broadening effect by turbulent diffusion. In a packed column, the average speed of a component is different in different zones, leading to various elution times (Figure 1.6). A can be expressed as $A = \lambda d$, where d is the average particle size of the packed filler, and λ represents the irregularity in packing. A depends on the geometry and uniformity of packing, but it is not affected by the amount of the stationary phase. A corresponds to the intercept in Figure 1.5. In a hollow capillary column, turbulent diffusion can be neglected.

Once a sample "pulse" is injected into a column, the narrow band will expand along the column under the driving force of the concentration gradient. Figure 1.7 shows the effect of this longitudinal diffusion on the elution of the samples through the column. In Equation 1.11, B is the longitudinal diffusion coefficient. $B = 2\gamma D_M$, which is related to the diffusion coefficient D_M of the sample in the carrier gas and the curvature factor γ of the filler particles. For a packed column, γ is approximately 0.6; for a capillary column, γ is 1. Longitudinal diffusion is inversely proportional to the carrier gas flow rate: its effect is more profound at a slow carrier gas flow rate; hence, the sample resides for a long time in the column. When $\bar{u}$ is

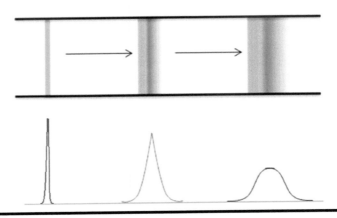

Figure 1.7   Longitudinal diffusion of a sample along the column.

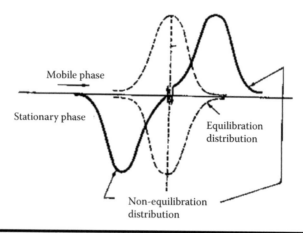

Figure 1.8   Non-equilibration caused by mass-transfer resistance.

small, longitudinal diffusion significantly contributes to peak broadening (shown in Figure 1.5).

The non-instantaneous equilibration of a component also contributes to peak broadening. In an ideal column, instantaneous equilibration is expected between the stationary and mobile phases. However, in an actual column, because of the finite mass-transfer rate, the part of a component that diffuses from the mobile phase into the stationary phase and then back into the mobile phase lags behind the other part that remains in the mobile phase. Hence, the peak is broadened (shown in Figure 1.8). C in Equation 1.11 represents the mass-transfer resistance; its effect is significantly more severe at a high carrier gas flow rate ($\bar{u}$), and C corresponds to the slope of the linear part in Figure 1.5.

In this section, only chromatography performed under isothermal operating conditions is discussed, which are suitable for a sample containing few components with a narrow range of boiling points. When a sample is composed of tens or even hundreds of components, and the boiling point range includes temperatures of greater than 80–100°C, chromatographic analysis is often carried out under temperature-programmed conditions. Thus, nearly every component is eluted at its optimal temperature, with a good resolution and symmetrical shape.

1.2.2 Instrument and Sample

GC is a simple instrument. Figure 1.9 shows a schematic of a typical GC instrument, comprising four parts: a carrier gas system, a separation system, a detection system, and a supplementary system. A vapor mixture is separated and eluted

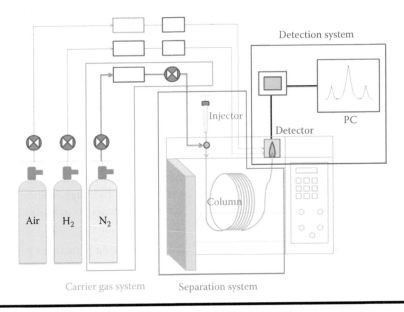

Figure 1.9 **Schematic of a gas chromatography instrument (supplementary system is not shown here).**

through a column using a carrier gas. Every component is detected, and a chromatogram is obtained.

1. Carrier Gas System

The carrier gas system comprises a gas circuit and a flow controller. The carrier gas is responsible for moving the mixture through a column until all components reach the detector. The carrier gas should be inert and highly pure; furthermore, it should not interact with the components and stationary phase. Typically, nitrogen, helium, and hydrogen are used as the carrier gases. Different detectors require the use of a specific carrier gas.

2. Separation System

The separation system comprises a sample injection unit and column. A liquid mixture is instantaneously vaporized in the injection unit, which is then passed into the column by the carrier gas for separation. The sample should be vaporized, but not decomposed, at the gasification temperature. Typically, liquid and gas samples of 0.5–10 μL and less than 10 mL, respectively, are required for GC analysis.

The column, which is often composed of glass or stainless steel, is the core part of a GC instrument. Two types of columns are available:

1. A packed column is produced by filling a column with a stationary phase; hence, the stationary phase can be easily changed. Typically, a packed column, which is often used for the analysis of not-so-complex mixtures, is 1–3 m

in length, with an inner diameter of 2–5 mm. This column can handle large sample loading, albeit it exhibits a low separation efficiency.
2. A capillary column is produced by coating, bonding, or crosslinking a stationary liquid onto the inner wall of a hollow column. Typically, a capillary column, which is often used for the analysis of complex mixtures with tens or even hundreds of components, is 1–100 m in length, with an inner diameter of 0.1–1 mm. This column exhibits a high separation efficiency, albeit can only handle low sample loading.

The column is the core part of a GC instrument, and the stationary phase is the essential part of the column. An appropriate stationary phase significantly affects the efficient separation of a mixture. Several stationary phases are available:

1. Absorbent: Typically, adsorbents are used for analyzing permanent gases and some low-boiling-point materials.
2. Stationary liquid: Typically, a stationary liquid is coated on inert porous supporting materials; it must be capable of dissolving all components and be completely nonvolatile under the operating range of temperature. In addition, it must be chemically and thermally stable, and exhibit a high boiling point. Various types of stationary liquids are available for use as the stationary phase.
3. Porous polymer microspheres: Typically, they are copolymers of styrene and divinylbenzene, which can be used individually or by coating stationary liquids on them.

How to select an appropriate stationary phase? There are no fundamental rules for selection. Its selection is dependent on its compatibility with (and selectivity for) components. When a stationary phase exhibits a chemical structure and polarity similar to those of certain components, it is believed to exhibit good selectivity. Typically, the following four stationary phases are suitable for almost all types of laboratory-scale separation experiments.

1. Nonpolar column: The stationary phase is polymethylsiloxane.
2. Weak-polar column: The stationary phase is poly(methylphenylsiloxane) with 5% phenyl groups.
3. Polar column: The stationary phase is polyethylene glycol with an average molecular weight of 20,000 (PEG-20M).
4. Al_2O_3 column: It is suitable for the analysis of hydrocarbons with less than six carbon atoms.

3. Detection System

The detection system is employed to detect the concentrations (or masses) of the eluted components with time. Typically, four detectors are used:

1. Thermal conductivity detector (TCD): TCD can measure the difference of the thermal conductivity between the carrier gas and component; hence, output signals are generated for all materials. It is a common detector with good stability and quantitative precision. However, its sensitivity is not as high as those of other detectors. Typically, hydrogen and helium are used as the carrier gas for TCD.
2. Flame ionization detector (FID): Eluted organic components are burnt in a hydrogen flame, ions are formed, and electronic signals are generated under an electric field, which are recorded. FID is a selective detector and does not detect signals for nitrogen, argon, helium, H_2O, CO, CO_2, NO, SO_2, and H_2S. Because of its high sensitivity (several magnitudes greater than TCD), FID is most widely utilized. Typically, nitrogen is used as the carrier gas for FID.
3. Electron capture detector (ECD). ECD is a selective detector with high sensitivity; it only responds to components with electronegative elements. Hence, it is suitable for materials containing halogen, sulfur, phosphorus, nitrogen, and oxygen.
4. Flame photometric detector (FPD). FPD is a selective detector, suitable for detecting sulfur- and phosphorus-containing compounds. Its response to sulfur and phosphorus is 10^4 greater than that to hydrocarbons.

Table 1.1 summarizes the typical performance indices of these detectors.

4. Supplementary System

The supplementary system comprises a temperature control unit and data processing unit. The temperatures in the sample injection unit, column box, and

Table 1.1   Typical Performance Indices of Detectors

Detector	Sensitivity	Linearity Range
Thermal conductivity detector	10^{-6} mg/mL	10^4–10^5
Flame ionization detector	10^{-12} g/s	10^6–10^7
Electron capture detector	10^{-14} g/mL	10^2–10^4
Flame photometric detector	~10^{-11} g/s	~10^4

detector are all controlled by the temperature control unit. The column box can be operated under an isothermal mode or a multiprogrammable temperature mode for specific samples to ensure that the complete separation of components can be realized as soon as possible. Low-temperature operation helps to improve separation at the cost of a lengthy analysis time. Typically, a temperature-programmed operation is used for mixtures with several components and wide boiling point ranges; hence, various components are eluted at their optimal temperatures in short time with a good separation.

In summary, chromatographic conditions can be selected according to the following rules:

1. The stationary phase can be selected according to its compatibility with the components.
2. The optimal flow rate of the carrier gas, $\bar{u}_{opt}$, can be calculated according to the Van Deemter equation. At this rate, the plate height is the minimum. In practice, $1.1\,\bar{u}_{opt}$ is typically used to shorten the analysis time.
3. The column temperature should cover the boiling point range of the mixture. The applicability of the stationary phase and the sensitivity of the detector at this temperature should be considered.
4. The temperature of the sample injector is set to the boiling point of the sample or at temperatures greater than 10–50°C.
5. Typically, less than 10 μL of a liquid sample is used, while less than 10 mL of a gas sample is used.

1.2.3 Data Processing

After a sample is injected into a GC instrument, its components are eluted and individually detected. A chromatogram is then obtained (Figure 1.10). The x- and y-axes represent the retention time and signal intensity, respectively.

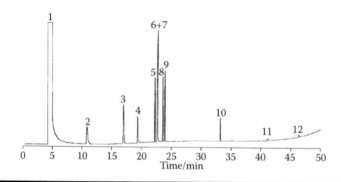

Figure 1.10   Typical chromatogram.

Table 1.2 Typical Definitions for Retention Parameters Used in GC

Name	Definition	Equation	
Retention time t_R (retention volume V_R)	The time (volume) from sample injection to the peak top of a component	$V_R = t_R F_c$	(1.12)
Dead time t_M (dead volume V_M)	The retention time (retention volume) of an inert component	$V_M = t_M F_c$	(1.13)
Adjusted retention time t'_R (adjusted retention volume V'_R)	The retention of a component in the stationary phase	$t'_R = t_R - t_M$ $V'_R = V_R - V_M$	(1.14)
Corrected retention time t^0_R (corrected retention volume V^0_R)	The retention time (volume) corrected by the correction factor of pressure gradient j	$t^0_R = j t_R$ $V^0_R = j V_R$	(1.15)
Net retention time t_N (net retention volume V_N)	Adjusted retention time (volume) corrected by j	$t_N = j t'_R$ $V_N = j V'_R$	(1.16)

The retention value can either be retention time or retention volume. Table 1.2 summarizes some of the typical definitions for retention parameters used in GC.

In Table 1.2, F_c (mL/min) is the flow rate of the carrier gas in the column at the column temperature, which can be calculated by the following equation:

$$F_c = F_0 (T_c / T_a).$$ (1.17)

Here, F_0 is the volume flow rate of the carrier gas at the exit temperature and pressure. T_c and T_a are the column temperature and room temperature, respectively (K).

The correction factor of the pressure gradient j is used to correct the pressure gradient in the column resulting from the compressibility of the mobile phase:

$$j = \frac{3}{2} \left[\frac{(P_i/P_0)^2 - 1}{(P_i/P_0)^3 - 1} \right].$$ (1.18)

Here, P_i and P_0 are the inlet and outlet pressures of the column, respectively (MPa).

1.2.3.1 Qualitative Analysis

GC is an efficient and rapid technique for the separation of mixed components. However, components cannot be identified by GC.

The qualitative identification of GC is based on the retention time. Under constant chromatographic conditions, the retention time of a component is a constant. Hence, qualitative analysis appears to be straightforward in that the unknown sample can be identified by the mere comparison of its retention time with that of a reference. If components do not exhibit the same retention time, the components are definitely not the same. However, if these components exhibit the same retention time, they may or may not be the same component. As the retention time significantly depends on the instrument and chromatographic conditions, it is not a sufficient and mandatory parameter for qualitative analysis.

Hence, for qualitative analysis, a parameter more appropriate than retention time is the retention index. The retention index of an n-alkane, I, is defined as follows:

$$I = 100 \times n. \tag{1.19}$$

Here, n is the number of carbons in n-alkane.

Under isothermal conditions, I of a component is expressed as follows:

$$I_x = 100 \left[z + \frac{\lg t'_{R(x)} - \lg t'_{R(z)}}{\lg t'_{R(z+1)} - \lg t'_{R(z)}} \right] \tag{1.20}$$

$$t'_{R(z)} \leq t'_{R(x)} \leq t'_{R(z+1)}$$

Here $t'_{R(z)}$ and $t'_{R(z+1)}$ are adjusted retention times of alkanes with the carbon numbers of z and $z+1$.

Under temperature-programmed conditions, I of a component is proportional to the number of carbons.

$$I_x = 100 \times \left[z + \left(t'_{R(x)} - t'_{R(z)} \right) / \left(t'_{R(z+1)} - t'_{R(z)} \right) \right] \tag{1.21}$$

Textbooks or published studies summarize the retention indices of several standard materials. Notably, the chromatographic condition of the component must be the same as that of n-alkanes.

In addition, some useful empirical rules must be followed:

1. Carbon number rule of homologs: Under isothermal conditions, the logarithm of the adjusted retention time of homologs is proportional to the numbers of carbons in a molecule.

$$\lg t'_R = an + b. \tag{1.22}$$

Here, a and b are constants.

2. Boiling point rule of isomers: The logarithm of the adjusted retention time of isomers is proportional to the boiling points:

$$\lg t'_R = cT_b + d. \tag{1.23}$$

Here, T_b is the boiling point, and c and d are constants.

Another method of qualitative analysis typically employed involves the combination of GC with other analytical techniques, such as mass spectrometry (MS) and infrared spectroscopy (IR). On the one hand, complex mixtures can be effectively separated by GC. On the other hand, each component can be effectively identified by MS or IR. Hence, currently, GC–MS or GC–IR is the most widely used combination.

1.2.3.2 Quantitative Analysis

GC is a powerful method for quantitative analysis, exhibiting precision and convenience. The peak area is proportional to the component concentration. As a detector produces different signal intensities for different materials, two components with the same concentration may give rise to different signals. Hence, sensitivity must be introduced for quantitative analysis.

The absolute sensitivity f_i is defined as the ratio of the weight of a component m_i to its peak area A_i:

$$f_i = \frac{m_i}{A_i}. \tag{1.24}$$

The frequently used parameter is the relative sensitivity f_{is}, i.e., the ratio of the absolute sensitivities of a component i to the standard s:

$$f_{is} = \frac{f_i}{f_s} = \frac{A_s m_i}{A_i m_s}. \tag{1.25}$$

With relative sensitivity, a sample can be quantitatively analyzed by any of the following methods, which are applicable for various cases.

1. Normalization: When all components in a sample exhibit chromatographic peaks, and their relative sensitivities are known, the weight of the component i can be calculated as follows:

$$x_i = \frac{f_{is} A_i}{\sum\limits_{i=1}^{n} f_{is} A_i} \times 100\%. \tag{1.26}$$

2. Internal standard: When some of the components in a sample cannot be eluted, or they do not exhibit chromatographic peaks, a known pure substance (internal standard) with a known weight can be added into the sample; this standard must be chosen such that it does not exhibit a peak that overlaps with any other component peak. For a component *i*, its relative sensitivity to the standard must be first measured, and its weight can then be calculated as follows:

$$x_i = \frac{m_s A_i f_{is}}{mA_s} \times 100\%. \tag{1.27}$$

Here, *m* and m_s are the masses of the sample and standard, respectively.

3. External standard: The same amount of a sample and a mixture of a pure substance (standard) and the component *i* are analyzed by GC under the same conditions. The weight of the component *i* can be calculated as follows:

$$x_i = E_i \frac{A_i}{A_E}. \tag{1.28}$$

Here, E_i and A_E are the weight and peak area of the component *i* in the standard, respectively.

4. Calibration curve: First, a series of standard samples with known weights of the component of interest are prepared. Then, the chromatographic analysis of the same amount of the standard samples under certain conditions is carried out. The peak areas of the component versus its concentrations are plotted, affording a calibration curve. The concentration of an unknown component in a sample can be analyzed under the same conditions, and the weight of the component can be calculated from the calibration curve from its peak area.

1.2.4 Applications of GC

For GC analysis, the sample must be a gas or a volatile liquid. Hence, GC cannot be used to directly analyze polymer materials. Monomers, solvents, or small-molecular-weight additives in polymer materials can be analyzed by GC. In addition, the change in the monomer concentration during polymerization can be measured to study polymerization kinetics. In combination with other techniques, such as pyrolysis described in Chapter 2, GC can be used to analyze polymer materials.

For analyzing the residual monomers or additives in polymer materials, low-boiling-point solvents can be used to extract these small molecules, followed by solution analysis with GC. If it is difficult to find an appropriate solvent, a head-space technique or solid microextraction for GC can be considered.

1.3 Inverse Gas Chromatography

1.3.1 Theory

A known stationary phase is used in typical GC analysis. An unknown sample is injected into the GC instrument, and various components are separated on the basis of their different interactions with the stationary phase. Conversely, it is also possible to use an unknown polymer as the stationary phase and inject a known substance (the probe) into the GC instrument. The change of the retention time (volume) of the probe represents the change of its interaction with the polymer. This technique was proposed by T. C. Davis in 1966 and is known as inverse gas chromatography (IGC).

In IGC, the retention time of a probe comprises three parts (shown in Figure 1.11): adsorption on the polymer surface; dissolution of, and diffusion into, the polymer coating; and further adsorption on the support surface. Under different conditions (e.g., at elevated temperatures), the polymer coating may be in the glassy, rubbery, or viscous state, and the interaction between the polymer coating and probe changes. The retention of the probe (specific retention volume V_g defined by Equation 1.29) changes accordingly. Hence, it is possible to investigate the physical changes of polymers and the interaction between polymers and small molecules via the change in the specific retention volume of the probe.

$$V_g = \frac{273}{T} \frac{V_N}{m_c}.$$

(1.29)

Here, T is the column temperature (K), and m_c is the mass of the stationary phase (polymer coating) (g).

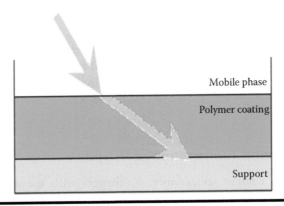

Figure 1.11 Adsorption of the probe on the surface and its diffusion into the polymer coating.

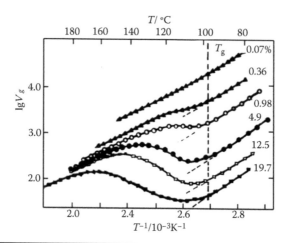

Figure 1.12 Effect of coating thickness on the retention of n-hexadecane in polystyrene.

1.3.2 Experimental

Typically, the stationary phase can be prepared by dissolving a polymer sample in a solvent and coating it on an inert support material (e.g., glass beads). Then, this stationary phase is filled into a column. In some cases, polymer films, fibers, or powders can also be filled into a column and used as the stationary phase.

During the preparation of a polymer coating, the coating thickness is crucial for subsequent analysis. Figure 1.12 shows lgV_g versus $1/T$ curves for different thicknesses of coatings using n-hexadecane and polystyrene as the probe and stationary phase, respectively. With increasing coating concentration (wt% to the support) from 0.079% to 19.7%, the curvature (glass transition) becomes significant.

1.3.3 Applications of IGC

IGC is convenient for detecting the transition temperatures and investigating the crystallinity and crystallization kinetics of polymers. In addition, the diffusion coefficients and activation energies of small molecules in polymers and interactions between them, as well as interactions between polymers and probes, can be investigated.

1.3.3.1 Transition Temperature of Polymers

With increasing column temperature, V_g changes with the different states of polymers. Figure 1.13 shows a typical "Z-shaped curve" for lgV_g versus $1/T$ for a semicrystalline polymer. The transitions in the curve correspond to the polymer transition temperatures. In the segment AB, the polymer is in the glassy state;

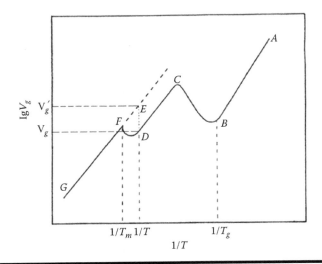

Figure 1.13 $\lg V_g$ versus $1/T$ curve for a semicrystalline polymer. (Note: B, other than at the bottom in the BC segment, corresponds to T_g.)

hence, the probe is only absorbed on the polymer coating surface. With increasing column temperature, desorption is accelerated, and adsorption becomes weak. Hence, V_g decreases accordingly. B represents the point at which the curve starts to turn upwards, and the temperature at B corresponds to the glass transition temperature of the polymer T_g. In the segment BD, the polymer changes from the glassy state to the rubbery state. In the BC segment, the polymer undergoes glass transition; hence, the probe starts to diffuse into the interior of the polymer. With increasing temperature, more and more polymer segments unfreeze from the glassy state. Thus, the probe can further diffuse into the polymer and reside there for a long time. Hence, V_g increases. With increasing temperature to C, the glass transition is completed. After this point, in the CD segment, the polymer is in the rubbery state. With an increase in temperature, the diffusion of the probe into and out of the polymer is accelerated; hence, V_g decreases again. In the DF segment, the crystal regions in the polymer begin to melt, and the probe diffuses into more amorphous regions. With increasing temperature, V_g increases until it reaches F, where the polymer is completely melted. F corresponds to the melting point T_m of the polymer. In the FG segment, the polymer is in the viscous state. With an increase in temperature, the dissolution and diffusion of the probe are accelerated. Hence, V_g decreases again.

In summary, in the AB segment, the probe is adsorbed onto the polymer coating surface. In the CD segment, the probe dissolves and only diffuses in the amorphous region of the polymer. In the FG segment, as the crystal regions melt, the probe dissolves and diffuses in the entire polymer. Hence, by plotting the $\lg V_g$ versus $1/T$ curve, T_g and T_m can be measured. Because the adsorption enthalpy is different from dissolution enthalpy, the slopes of these segments are different. For an amorphous polymer, the DFG segment does not exist because there is no crystal region. If the crystallinity of a polymer is extremely high, the glass transition region (the ABC segment) may not be obvious or even disappear.

1.3.3.2 Crystallinity and Crystallization Kinetics of Polymers

In Figure 1.13, the specific retention volume at a temperature T is V_g (at point D). When the segment GF is extrapolated to low temperature, it meets the temperature T at point E, which corresponds to V_g'. Hence, the crystallinity X_c of a semicrystalline polymer at T can be calculated from their ratio:

$$X_c = 1 - \frac{V_g}{V_g'}.$$ (1.30)

For the measurement of crystallinity by IGC, the parameters of the crystal and amorphous regions (e.g., specific volume), as well as the polymer mass and carrier gas flow rate, are not required. X_c can be calculated by the ratio of V_g and V_g', which can be easily measured from IGC analysis. Hence, IGC is clearly advantageous over the density and x-ray methods for the determination of the crystallinity of a new polymer.

In addition, IGC can also be used to investigate the crystallization kinetics of a polymer. First, the column temperature should be increased to some temperature greater than T_m. Next, sufficient time should be provided for the complete melting of the polymer at this temperature. Then, the column temperature should be decreased to some temperature less than T_m, and the change of V_g with time should be determined. The decrease rate of V_g corresponds to the growth rate of crystals. As V_g', at this temperature does not change with time, the change of X_c with time can be obtained, which corresponds to the kinetics curve for isothermal crystallization.

1.3.3.3 Diffusion Coefficient and Activation Energy of Small Molecules in Polymers

By studying the diffusion behavior of small molecules in polymers, the permeation performance of small molecules in polymer membranes or migration of small-molecular-weight additives in polymers can be studied.

The coefficient of the mass-transfer resistance C in Equation 1.11 can be obtained from the slope of the linear curve at high flow rates in Figure 1.5, which shows the plate height at different carrier gas flow rates. From principles of chemical engineering, for spherical support material particles, C can be expressed as follows:

$$C = \frac{8}{\pi^2} \frac{d_f^2}{D_l} \frac{K'}{(1+K')^2}. \tag{1.31}$$

Here, d_f is the thickness of the polymer coating on the support particle surface, and D_l is the diffusion coefficient of the probe in the polymer. K' is defined as follows:

$$K' = \frac{t_{R'}}{t_M}. \tag{1.32}$$

Hence, D_l can be calculated from Equation 1.31.

If the relationship between D_l and temperature follows the Arrhenius equation, the activation energy for diffusion ΔE_D can be obtained from Equation 1.33 by obtaining D_l at various temperatures and plotting $\lg D_l$ versus $1/T$:

$$\lg D_l = \lg D_l^0 - \Delta E_D / RT \tag{1.33}$$

Reference

1. Peizhang Lu and Chaozheng Dai. *Basic Principle of Chromatography*. Beijing: Science Press, 1989.

Exercises

1. Please explain the separation efficiency and resolution of a chromatographic column. If two components exhibit different retention times, can they be distinguished in a chromatogram?
2. Among retention time, dead time, and adjusted retention time, which parameter more appropriately reflects the interaction of components with the stationary phase? Why?
3. How many parts are present in a GC apparatus? What are their functions?
4. How is qualitative analysis conducted by GC?

5. How is quantitative analysis conducted by GC? What is the prerequisite for each method?

6. What is absolute or relative sensitivity? Why are these parameters required for quantitative analysis?

7. Which detector is suitable for the analysis of a mixture containing CO, H_2S, H_2O, and H_2? How is the carrier gas selected for its separation?

8. What is the difference between GC and IGC?

9. GC analysis was carried out under the following conditions: room temperature, 25°C; atmosphere pressure, 0.101325 MPa; gauge pressure at the column inlet, 0.066 MPa; column length, 2 m; column temperature, 50°C; and carrier gas flow rate in the column, 18.4 mL/min. Data for the components are listed in the following table:

Sample	t_R/min	W/min	Sample	T_R/min	W/mm
Air	0.28		Cyclohexane	7.22	0.640
Propane	1.57	0.140	n-Hexane	7.89	0.705
n-Butane	2.49	0.225	n-Propanol	8.88	0.800
2-Butane	2.95	0.265	n-Butanol	10.11	0.910
n-Pentane	4.38	0.395	n-Heptane	14.36	1.285

Please calculate:

(1) The correction factor of pressure gradient j
(2) The dead time of n-pentane
(3) V_R and V_N of n-pentane
(4) The theoretical plate number n and effective plate height H_{eff} based on cyclohexane
(5) The retention indices of 2-butane, n-hexane, and n-butanol
(6) The adjusted retention time of ethanol
(7) And the resolution of cyclohexane and n-hexane. Are they completely separated?

If all components have the same concentration, do they have the same peak areas? Why?

10. The theoretical plate numbers of a 1 m column at carrier gas flow rates of 10, 20, and 40 mL/min are 1205, 1250, and 1000, respectively. Please calculate A, B, and C in the Van Deemter equation and the optimal flow rate of the carrier gas and plate height in this case.

11. For the GC analysis of some aromatic isomers, 1.5 g of a mixture with a known weight was prepared first. Then, 1.5 mg of a known mixture and an unknown sample were analyzed. The results are listed in the following table:

Component	ethylbenzene	p-xylene	m-xylene	o-xylene
Mass in the known mixture/g	0.4051	0.3528	0.2834	0.4587
Peak area in the known mixture/mm²	116.0	98.0	82.0	130.0
Peak area in the unknown sample/mm²	138.0	86.0	166.0	115.0

Please calculate the absolute sensitivity of each component and their concentrations in the unknown sample.

12. The glass transition temperature of polystyrene (PS) was determined by IGC. Results are shown in the following table. Please plot the $\lg V_g$ versus $1/T$ curve and report the T_g value of PS.

Column Temperature/°C	Retention Time/s	Dead Time/s	Flow Rate of Carrier Gas in the Column/(mL/min)
80	72.2	27.2	19.3
90	55.9	28.0	20.3
95	52.1	28.2	21.3
100	51.6	28.4	23.2
105	52.7	28.2	23.9
110	52.7	29.1	24.4
120	52.5	29.6	23.4

13. For the IGC analysis of amorphous polystyrene and semicrystalline polypropylene, please indicate the change of the specific retention volume of a probe for both and explain the reasons.

Chapter 2

Pyrolysis Gas Chromatography and Mass Spectrometry

Pyrolysis gas chromatography (PGC) involves the use of a pyrolyzer and a gas chromatography (GC) instrument. By using the pyrolyzer, which is set at a specific temperature or operated according to a temperature program, a long polymer chain is rapidly cleaved into small molecular fragments (pyrolysates) under an inert atmosphere. The so-formed fragments are passed through a GC system for separation and detection. Under specific conditions, a polymer chain is cleaved in a specific manner, affording characteristic pyrolysates. Hence, as an analytical method, PGC can effectively identify a polymer via the analysis of its pyrolysates, similar to a jigsaw puzzle.

2.1 Pyrolysis Mechanism of Polymers

2.1.1 Pyrolysis

The pyrolysis of a polymer chain is carried out at 400–900°C, which can be regarded as the opposite process of polymerization:

1. Initiation: A hydrogen atom is lost from the chain or the chain is cleaved into two parts, affording corresponding radicals.

$$M_n \rightarrow M_n \cdot$$

$$M_n \rightarrow M_i \cdot + \cdot M_{n-i}$$

2. Propagation or transfer: Depolymerization, or chain transfer, results in the formation of monomers or oligomers, respectively.

$$M_i \cdot \rightarrow M_{i-1} \cdot + M$$

$$M_i \cdot \rightarrow M_{i-2} \cdot + M_2$$

$$M_i \cdot \rightarrow M_{i-n} \cdot + M_n$$

3. Termination: Coupling, or disproportionation, results in the disappearance of radicals.

$$M_i \cdot + M_k \cdot \rightarrow M_{i+k}$$

$$M_i \cdot + M_i \cdot \rightarrow M_{2i-1} + M$$

The ability of a polymer to undergo pyrolysis is evaluated by zip length (ZL) [1]. ZL corresponds to the number of monomers formed from a macroradical (Figure 2.1): the change of polymerization degree (DP/DP_0) is plotted against polymer weight loss, where DP and DP_0 represent the average polymerization degree and original polymerization degree, respectively. Curve *a* represents the depolymerization of a polymer chain. Large amounts of monomers are produced by pyrolysis. Hence, the mass of the polymer is rapidly lost, while its molecular weight slightly decreases

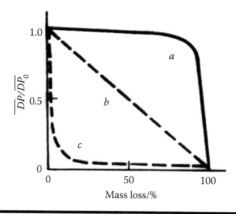

Figure 2.1 **Change of the average polymerization degree versus the mass loss of polymers.**

at the start. In this case, ZL is high, e.g., the ZL of poly(methylmethacrylate) (PMMA) is 2.5×10^3. Curve *c* represents the random scission of a polymer chain. In this case, the molecular weight of the polymer rapidly decreases. However, as small volatile fragments are produced, the mass marginally decreases at the start. In this case, ZL is near zero, e.g., polyethylene (PE). The pyrolysis behavior of curve *b* is intermediate between those of *a* and *c*, showing a part of depolymerization and part of random scission, with the simultaneous decrease of the molecular weight and mass of the polymer. For example, the ZL of polystyrene (PS) is 3.

2.1.2 Pyrolysis Patterns

Under specific conditions, the pyrolysis behavior of a polymer chain, as well as the pyrolysates obtained, is unique and depends on the polymer structure. Hence, it is crucial to obtain information about the pyrolysis patterns [1] of different polymers for the identification and analysis of their structures and fragments.

1. Random Scission of the Main Chain
    The pyrolysis of vinyl polymers occurs via the random scission of the main chain, affording a series of fragments with different polymerization degrees. The monomer yield is quite low, and ZL is near zero.

For example, the pyrolysis of PE occurs by random scission, where the pyrolysates formed are mainly a series of n-alkenes with different numbers of carbon atoms (Figure 2.2).

For substituted vinyl polymers, the presence of tertiary atoms permits facile chain scission at such branching points, resulting in corresponding fragments of high intensity. For example, in the pyrogram of polypropylene (PP), higher-intensity peaks are observed for trimer, tetramer, and pentamer (Figure 2.3).

2. Main Chain Scission Induced by the Elimination of Substituents
    The pyrolysis of polyvinyl chloride (PVC) is a typical example, where HCl is eliminated by the following reaction, and a C=C bond is formed in the main chain:

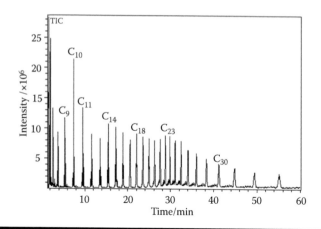

Figure 2.2 Pyrogram of PE.

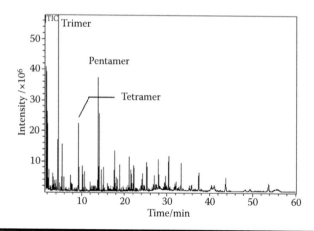

Figure 2.3 Pyrogram of PP.

The elimination of HCl continues, resulting in the formation of a conjugated alkene structure in the main chain:

$$\text{wwww } C-C-C=C-C \text{ wwww } \longrightarrow \text{ wwww } C=C-C=C-C \text{ wwww } + HCl$$
$$\qquad\ |$$
$$\qquad Cl$$

At a set temperature of 250–350°C, polyyne is generated. At around 500°C, further chain scission rapidly occurs, following elimination.

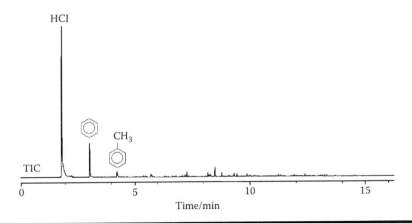

Figure 2.4 Pyrogram of PVC. (Reprinted from *Pyrolysis-GC/MS Data Book of Synthetic Polymers*, Tsuge, S., H. Ohtani, and C. Watanabe, Copyright 2011, with permission from Elsevier.)

Radical transfer via a six-membered-ring transition state occurs, leading to the formation of benzene as the main product (Figure 2.4) [2].

3. Depolymerization

In case of an α-substituted polymer, terminal radicals can be rapidly formed and decomposed via unzipping. The monomer is the main pyrolysate obtained, with a high ZL, e.g., PMMA undergoes pyrolysis by depolymerization, and its pyrogram is shown in Figure 2.5 [2].

MMA

TIC

0   5   10   15

Time/min

Figure 2.5 Pyrogram of PMMA. (Reprinted from *Pyrolysis-GC/MS Data Book of Synthetic Polymers*, Tsuge, S., H. Ohtani, and C. Watanabe, Copyright 2011, with permission from Elsevier.)

4. Intramolecular Cyclization

The intramolecular cyclization of polyacrylonitrile (PAN) is a typical example. At approximately 200°C, PAN can change from a linear structure to a ladder-like structure via intramolecular cyclization:

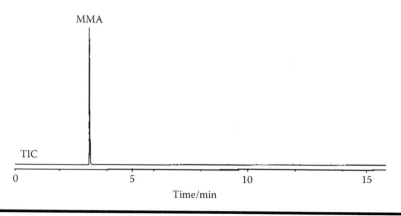

On the other hand, if PAN is heated to 500–600°C, the chain is rapidly cleaved, affording large amounts of oligomers and monomers.

If hydrogen atoms on the tertiary carbon atoms are replaced by methyl groups, the so-formed polymethyl acrylonitrile is prone to depolymerize to its corresponding monomers instead of intramolecular cyclization.

5. Pyrolysis of Polymers with Unsaturated Main Chains

Polymers such as polybutadiene (PB), with unsaturated bonds in the main chain, are easily cleaved at the α or the β position:

β-scission affords the 1,3-butadiene monomer, while α-scission affords the vinyl cyclohexene dimer via radical transfer. At 500°C, the main pyrolysates are monomers and dimers.

6. Pyrolysis of Polymers with Other Atoms in the Main Chain

In the presence of atoms, such as nitrogen and oxygen, in addition to carbon atoms in the polymer main chain, the bond energy of C–X (X=N, O, and others) is typically less than that of C–C. Hence, polymers are easily cleaved at these weak points, i.e., at α and β positions, as follows:

Polyamide

Polyester

Polyether

Polycarbonate

2.2 Pyrolysis Gas Chromatography

A PGC system can be easily set up by introducing a pyrolyzer on the top of the injection port of a typical GC instrument.

- PGC can be widely employed in various applications, such as in the petrochemical, pharmaceutical, biomedical, forensic, archaeological, geological, coal industry, and the environmental industry, without any limitations of the state and shape of samples.
- Pyrolysis conditions can be easily set to obtain structural information from different aspects or to investigate the behavior of samples under simulated processing or service conditions. PGC is sensitive to subtle structural differences; thus, it is powerful for investigating complex problems, such as stereoregularity and sequence distribution.
- Marginal amounts of samples (10–100 μg) are typically required for PGC analysis. Pre-purification or complex pretreatment is not required.

Furthermore, solvents, additives, and polymers can be separately analyzed by a flexible multistep procedure. PGC is advantageous for analyzing thermosets, which are insoluble and cannot melt, hence, the analysis of thermosets by several other analytical methods is difficult.

■ As pyrolysis temperature significantly affects pyrolysates, and secondary reactions of radicals possibly occur in a pyrolyzer, pyrolysis is a complex process. In addition, issues associated with marginal sample amounts, caused by the heterogeneous nature of polymer materials, are observed.

2.2.1 Pyrolyzer and Pyrolysis Conditions

For obtaining reproducible and characteristic pyrograms, a pyrolyzer must satisfy the following requirements:

■ Sufficiently wide temperature range with good precision
■ Rapid heating speed
■ Small dead volume to prevent secondary reactions of fragments
■ No catalysis effect

Figure 2.6 shows the three commonly utilized pyrolyzers.

(1) Filament Pyrolyzer

Typically, a platinum (Pt) filament or belt is used because it is inert and can be rapidly heated to the set temperature. Samples are coated on the filament or belt and pyrolyzed at high temperatures. The construction of this

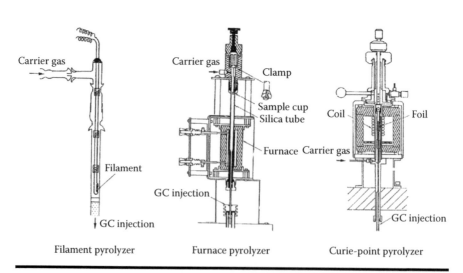

Figure 2.6 Three commonly utilized pyrolyzers.

pyrolyzer is simple, with a small dead volume. The heating rate ranges from 0.01°C/min to 30,000°C/min.

(2) Furnace Pyrolyzer

A sample added in a small stainless cup is sent into a silica tube with a surrounding furnace; this furnace can be heated to a set temperature or via a temperature program. The furnace pyrolyzer can handle samples of different shapes; hence, it can be employed in various applications. Nevertheless, a slightly high dead volume is observed, and secondary reactions are possible.

(3) Curie-point Pyrolyzer

Samples are enclosed using ferromagnetic foils and then placed in a radio-frequency induction coil. Its temperature rapidly increases before it reaches the Curie point. Then, heating is ceased, and the temperature remains constant. The pyrolysis temperature (Curie temperature) depends on the composition of the alloy (Table 2.1). The heating rate of this pyrolyzer is rapid, with a small dead volume, but the temperature cannot be freely set and programmed.

The pyrolysates are significantly affected by pyrolysis conditions. Typically, at low pyrolysis temperatures, a slow degradation rate is observed, characteristic fragments are not significant, and several byproducts and oily products with high boiling points are obtained. At high temperatures, the fragments are too small to be characteristic. For example, the main pyrolysates of PS at 425°C are styrene and dimer, while those at 825°C are styrene, benzene, toluene, ethylene, and ethyne. On the other hand, at 1025°C, large amounts of benzene and ethylene are produced, and these pyrolysates do not reflect the original polymer structure. Hence, it is crucial to optimize and precisely control the pyrolysis temperature so as to obtain reproducible and characteristic programs.

PGC is quite sensitive; hence, only marginal amounts of a sample are required (typically, 10–100 µg). However, the use of such marginal amounts proves to be

Table 2.1   Alloy Composition and the Corresponding Curie Temperature

Composition (%)	Curie Temperature/°C	Composition (%)	Curie Temperature/°C
Ni(100)	358	Ni(67):Co(33)	660
Fe(48):Ni(51):Cr(1)	440	Fe(100)	770
Fe(49):Ni(51)	510	Ni(40):Co(60)	900
Fe(40):Ni(60)	590	Fe(50):Co(50)	980
Fe(30):Ni(70)	610		

difficult for the analysis of polymer composites or blends, which are heterogeneous in nature; sampling from different positions of such samples can possibly lead to different compositions; hence, reproducibility is difficult. In this case, multianalysis should be carried out, and an average result should be reported.

2.2.2 Derivatization Techniques of PGC

1. Pyrolysis-hydrogenation Gas Chromatography (PHGC)

    Pyrolysates of vinyl polymers include a series of alkanes, alkenes, and dienes with different numbers of carbons as well as various branches. The pyrogram is complex with tens to hundreds of peaks. Hence, for simplifying the pyrogram and highlighting characteristic peaks, a hydrogenation column (using a hydrogenation catalyst) is introduced between the pyrolyzer and the injection port of a GC instrument. Alkenes, dienes, and other unsaturated fragments are saturated by hydrogen, affording corresponding alkanes. Hence, the number of peaks significantly decreases, and peak intensities increase. For example, the pyrolysis of PE affords n-decene and α,ω-decadiene, which change to n-decane after hydrogenation:

C-C-C-C-C-C-C-C-C-C
C=C-C-C-C-C-C-C-C-C $\xrightarrow{\text{H}_2}$ C-C-C-C-C-C-C-C-C-C
C=C-C-C-C-C-C-C=C

    Figure 2.7 shows the pyrograms of PE before and after hydrogenation [3]. PHGC can be utilized to investigate the branching and stereoregularity of polyolefins.

2. Thermally Assisted Hydrolysis and Alkylation (THA)

    During the analysis of polar polymers (e.g., polyethers, polyamides, epoxy resins, and phenol resins) and natural polymers (e.g., saccharides and amino acids), polar pyrolysates can be easily adsorbed onto the inner walls of the pyrolyzer and the column, thereby contaminating the instrument and decreasing peak resolution. Thermally assisted hydrolysis and alkylation (THA) can be utilized for solving this problem. With THA, samples are co-pyrolyzed with a derivatization reagent (generally, organic alkali agents, such as $(CH_3)_4NOH$ and $(C_4H_9)_4NOH$), and pyrolysates (e.g., acids, alcohols, amines, and phenols) react with the derivatization reagent at high temperatures, affording esters and ethers. By this technique, peak resolution is significantly improved, and the characteristic fragments of high intensity are obtained. In addition, pyrolysis can be conducted at a relatively low

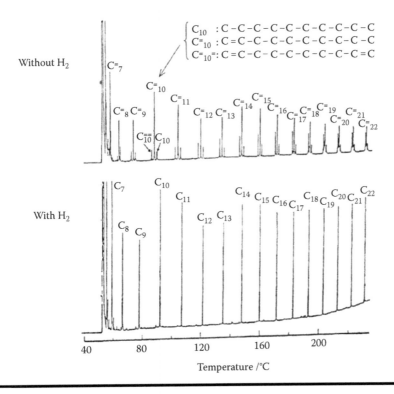

Figure 2.7 Pyrograms of PE before and after hydrogenation. (From Tsuge, S. et al., *Journal of Analytical and Applied Pyrolysis* 1, no. 3, (2000): 221–229.)

temperature with fewer residues. Figure 2.8 shows the pyrograms of polysulfone with and without THA [4]. A better pyrogram is obtained with THA, with characteristic peaks.

2.2.3 Pyrogram Interpretation

Pyrograms are similar to chromatograms, where peak intensity is plotted versus retention time. In a chromatogram, each peak represents a component in a mixture, while even in a pyrogram of a pure polymer, each peak only represents a polymer fragment, with all peaks completely representing the polymer. A more complex pyrogram is obtained for a composite or blend. Hence, a pyrogram is similar to a "fingerprint"; it is composed of characteristic peaks and needs to be interpreted as a whole.

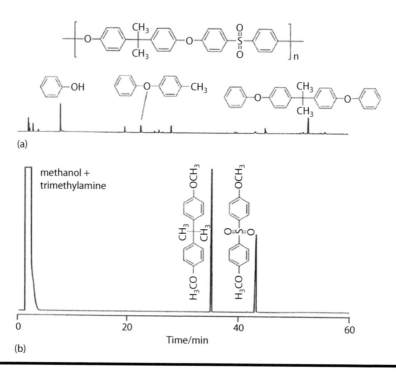

Figure 2.8 **Pyrograms of polysulfone with and without THA. (a) 700°C without (CH₃)₄NOH; (b) 300°C with (CH₃)₄NOH. (From Wampler, T.P. *Applied Pyrolysis Handbook* (2nd edition). CRC Press, Boca Raton, FL, 2007.)**

As described in Section 2.1.2, polymers exhibit characteristic pyrograms under specific pyrolysis conditions. Hence, polymers can be easily differentiated and identified by the comparison with standard or reference pyrograms [2]. In addition, the retention index (RI) (Section 1.2.3.1) can also be used for qualitative analysis. In GC, n-alkanes are used as references. In PGC, n-alkenes obtained from the pyrolysis of PE can be used as references [5]. Samples and PE are pyrolyzed under the same conditions so that RIs corresponding to the characteristic peaks (monomers and oligomers) of samples can be calculated according to Equation 1.21.

PGC is often coupled with a mass spectrometry (MS) instrument, furnishing PGC–MS, where MS functions as the detector for pyrolysates. PGC–MS combines the excellent separation ability of GC and good qualitative ability of MS. Currently, commercial PGC and PGC–MS standard libraries are available (e.g., F-Search of Frontier) for the search and qualitative analysis of unknown polymers.

2.3 Organic Mass Spectrometry

Organic MS, also known as molecular MS, is a powerful method for the qualitative analysis of organics. It is often coupled with GC or liquid chromatography (LC) and functions as a detector for effectively improving the weak qualitative ability of GC or LC.

2.3.1 Principle of MS

In MS, a sample is vaporized and bombarded by energetic electrons under high vacuum, affording positively charged ions; these ions are individually collected according to their mass to charge ratio (m/e), and a mass spectrum of ion intensity versus m/e is obtained. With specific electron energy, the formed fragment ions, as well as their intensities, depend on the sample molecule. Hence, MS is employed for structural analysis.

Various ions are observed in a mass spectrum, such as molecular ions, isotopic ions, fragment ions, multicharged ions, and metastable ions. These ions can be further categorized into odd-electron ions (OE^+) and even-electron ions (EE^+). When a molecule losses one electron, $OE^{+\cdot}$ is formed. $OE^{+\cdot}$ follows the "nitrogen rule," implying that the molecular weight of a compound without nitrogen atoms or with an even number of nitrogen atoms is an even number, while that with an odd number of nitrogen atoms is an odd number. $OE^{+\cdot}$ is an ion and a radical; hence, it is quite active and is easily cleaved further to smaller $OE^{+\cdot}$ or EE^+ fragments. EE^+ does not have unpaired electrons and does not follow the "nitrogen rule." When EE^+ is cleaved, smaller EE^+ fragments are formed.

1. Molecular Ion

    When a molecule loses an electron via the bombardment of energetic electrons, a molecular ion is formed, which is clearly $OE^{+\cdot}$.

$$M^{\cdot\cdot} + e \longrightarrow M^{+\cdot} + 2e$$
$$\text{neutral molecule} \qquad \text{molecular ion}$$

The mass of the molecular ion corresponds to the molecular weight of the compound. The intensity of the molecular ion depends on its stability. For compounds such as aromatics or compounds with a conjugated structure, the intensity of the molecular ion peak is high. For compounds containing -OH, $-NH_2$, and other groups (e.g., O, N, P, and S) or with substituents, the molecular ion peak is weak or even disappears. The order of the peak intensities of the molecular ion is as follows:

Aromatics > conjugated alkenes > alkenes > cyclic compounds

> carbonyl compounds > linear alkanes > ethers > esters

> amines > acids > alcohols > highly branched alkanes.

The peak intensity of the molecular ion is clearly related to the electron energy: peak intensity increases with decreasing electron energy.

The required, but not sufficient, conditions for the molecular ion are as follows:

(1) The molecular ion should exhibit the highest mass (including its isotopic ions).

(2) The molecular ion should be OE$^{+\cdot}$ and follow the "nitrogen rule."

(3) The molecular ion should have plausible fragment ions.

2. Isotopic Ions

Several elements have natural stable isotopes. Typically, the lightest isotope with a mass M exhibits the highest abundance, and the masses of heavier isotopes are $M+1$, $M+2$, and others. These isotopes contribute to the isotopic peaks in the mass spectra. Table 2.2 summarizes the common elements and isotopes in compounds and their abundances.

The intensity ratio of these isotopic peaks is related to the abundance of isotopes, which can be calculated according to the expansion of the following binomial:

$$(a + b)^n \tag{2.1}$$

Here, a is the abundance of the light isotope; b is the abundance of the heavy isotope; and n is the number of the concerned atom in a molecule. The number of items in the expansion of Equation 2.1 indicates the number of isotopic peaks, and the ratio of the items indicates the intensity ratio. Figure 2.9 shows the isotopic peaks of some halogen-containing compounds.

Table 2.2 Abundances of Common Elements and the Isotopes (Assuming That the Abundance of the Lightest Element Is 100)

Element	Lightest Isotope	M+1	Abundance	M+2	Abundance
Hydrogen	^{1}H	^{2}H	0.016		
Carbon	^{12}C	^{13}C	1.08		
Nitrogen	^{14}N	^{15}N	0.38		
Oxygen	^{16}O	^{17}O	0.04	^{18}O	0.20
Silicon	^{28}Si	^{29}Si	5.10	^{30}Si	3.35
Chlorine	^{35}Cl			^{37}Cl	32.5
Bromine	^{79}Br			^{81}Br	98.0

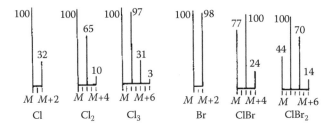

Figure 2.9 Intensity ratio of isotopic peaks for halogen-containing compounds.

For instance, $^{35}Cl:^{37}Cl \approx 3:1$. If there is only one chlorine atom in a molecule, two isotopic peaks are observed with an intensity ratio of $M:(M+2) \approx 3:1$. If there are two chlorine atoms, the binomial expansion is $a^2+2ab+b^2$. By substituting $a = 100$ and $b = 32.5$, the values of 10,000, 6500, and 1056.25 are obtained. Hence, the intensity ratio of these isotopic peaks is $M:(M+2):(M+4) = 10,000:6500:1056.25 \approx 9:6:1$.

Molecular ion isotopes are crucial for identifying the molecular formula. With the change in the elemental composition, the intensity ratio of isotopic peaks changes accordingly. These isotopes are also helpful for judging the presence of some special elements. If the M+2 intensity is less than 4% of the M intensity, Cl, Br, and Si are absent. If the M+1 peak is strong, several carbon atoms are often present. If this is not the case, Si may be present.

3. Fragment Ions

Typically, the ionization of organic molecules is carried out with 10–15 eV of electrons. Nevertheless, the electron energy for MS is typically 70 eV. The significantly higher energy leads to the dissociation of the molecular ion into various fragment ions. Hence, the molecular ion structure is deduced from the fragment ions.

4. Multicharged Ions

For a highly stable molecule, two or more electrons can be lost, affording m/2e or m/3e ions. This behavior is observed for aromatic and conjugated molecules.

5. Metastable Ions

If a parent ion m_1^+ is not stable, it can dissociate into m_2^+ when it passes from the ion source to the detector. In this case, m_1^+ may be accelerated in the ion source, while m_2^+ is possibly deflected in the mass analyzer. Finally, the determined m/e does not correspond to either m_1 or m_2, and the mass is possibly not an integer.

For brevity, only a common molecular ion, isotopic ions, and fragment ions are discussed in this book.

2.3.2 Mass Spectrometer

From the principle of MS, a mass spectrometer contains a sampling system, an ion source, a mass analyzer, a detector, and a set of vacuum systems.

1. Sampling System

   A sample is introduced by the sampling system into the mass spectrometer under high vacuum, followed by sample evaporation. It is imperative for the sample to be vaporized, not decomposed, in the sampling system.

2. Ion Source

   Two ion sources are commonly utilized: those from electron ionization (EI) and chemical ionization (CI). Sample molecules are bombarded with energetic electrons from the EI source, affording ions; the energy of electrons is typically 70 eV, significantly greater than the ionization energy of a molecule; hence, the molecular ion is further cleaved into fragment ions. For obtaining the molecular ion of a not-so-stable molecule, a CI source is a good choice. A reactive gas (typically CH_4) is introduced into the CI source, and the ions of the gas react with the sample molecule, affording the molecular ion, as shown below:

   $$A + HX^+ \rightarrow [A - H]^+ + X + H_2 \tag{2.2}$$

Or,

   $$A + HX^+ \rightarrow AH^+ + X \tag{2.3}$$

   Here, A is the sample molecule, and HX^+ is the gas ion. The product in Equation 2.2 is $(M-1)^+$, while that in Equation 2.3 is $(M+1)^+$.

   In addition to the commonly used EI and CI sources, field ionization (FI) and field desorption ionization (FD) sources are also utilized to obtain strong molecular ions. In an FI source, an electron from a molecule is directly "pulled out" under a very high electric field, affording a molecular ion. In an FD source, a sample is attached to a filament. When the current is switched on, the sample is desorbed, where it diffuses to a field-emission region, followed by ionization. The intensity of the molecular ion peak generated is even greater than that generated by an FI source.

3. Mass Analyzer

   The primary function of a mass analyzer is to separate ions with different m/e values. It is the core of a mass spectrometer.

   (1) Single-Focusing Mass Analyzer

   Figure 2.10 shows the mechanism of a single-focusing mass analyzer. After the ionization of a molecule, the ion is accelerated at the exit of the

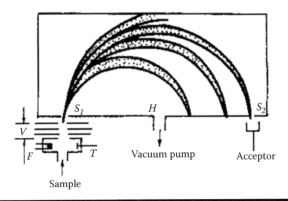

Figure 2.10 Schematic of a single-focusing mass analyzer.

ionization source by the electric field (assuming that its initial velocity is zero).

$$\frac{1}{2}mv^2 = eU \tag{2.4}$$

Here, m is the ion mass; v is the ion velocity; e is the charge quantity; and U is the voltage.

When the ion with a velocity v enters the mass analyzer, its movement is deflected in a curved path under the magnetic field. The centrifugal force balances the magnetic force:

$$evH = mv^2/r_m \tag{2.5}$$

Here, H is the strength of the magnetic field, and r_m is the radius of circular motion. From Equations 2.4 and 2.5:

$$m/e = \frac{H^2 r_m^2}{2U} \tag{2.6}$$

From Figure 2.10, if H and U are known, only ions with a radius of circular motion, r_m, can reach the acceptor and be detected. From Equation 2.6, if r_m is known, U can be scanned at a constant H, or H can be scanned at a constant U, for detecting ions in turn with increasing m/e. The former and latter processes are referred to voltage scanning and magnetic field scanning, respectively. In practice, a double-focusing mass analyzer is often used instead of a single-focusing mass analyzer for improving mass resolution and eliminating energy dispersion.

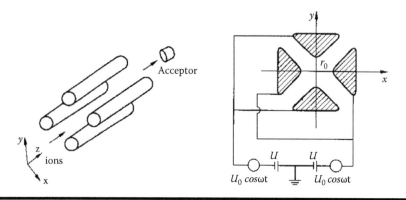

Figure 2.11   Schematic of a quadrupole mass filter.

(2) Quadrupole Mass Filter

A quadrupole mass filter is composed of four parallel electrodes with a hyperbolic cross section (Figure 2.11). One set comprises two electrodes at a diagonal, and two sets of electrodes are present. The same radio-frequency voltage $U_0 \cos(\omega t)$ and direct voltage U with opposite directions are applied to the two sets, and an electric field is formed among the quadrupoles. When ions from the ionization source enter into the electric field, these ions oscillate. Under a certain electric field, only an ion with a particular m/e exhibits stable oscillation and enters into the detector under the electric field, while other ions are neutralized by the electrodes. If U_0 is changed, but U_0/U is maintained constant (U will change accordingly), ions with different m/e can be scanned. This mass analyzer is advantageous for its scanning speed; thus, it is suitable for GC–MS.

4. Detector

A detector converts the detected cations into electrical signals, which are then amplified for subsequent data treatment.

5. Vacuum System

The entire system must be operated under a high vacuum, otherwise, a considerable number of nitrogen and oxygen atoms, as well as other gas molecules in the air, will also be bombarded by the ionization source. This, in turn, can generate a significant baseline noise or even bury the signals. Simultaneously, the filaments emitting energetic electrons may be severely oxidized and rapidly burnt out.

Typically, the vacuum system ensures that the entire system is under a high vacuum. In general, two pumps, turbo and molecular pumps, are used for maintaining low- and high-vacuum conditions, respectively.

When a mass spectrometer is connected to a GC with a capillary column, the outlet end of the column is directly introduced into the ionization source, because the flow of carriers in the column is low, and the pump power is sufficiently high.

2.3.3 Mass Spectral Interpretation

2.3.3.1 Mass Spectrum

Typically, a mass spectrum is similar to a bar graph; that is, a peak is shown as a vertical line. The peak height corresponds to the peak intensity (Figure 2.12). The X- and Y-axes represent m/e, showing the mass of the fragment ions, and ion intensity, respectively. Typically, the highest peak is defined as the base peak, with an intensity of 100%. The relative abundances of other peaks are expressed as their percentages to the base peak.

2.3.3.2 Typical Fragmentation

In the mass spectra of organic compounds, the majority of fragment ions are formed according to certain rules, and the fragmentation mode is closely related to the molecular structure and functional groups. Thus, only when the fragmentation nature is understood, qualitative analysis is possible. As mentioned in Section 2.3.2, molecular ions are predominantly produced by CI, FI, and FD, while several fragments are produced by EI, making it possible to examine the molecular structure. Hence, in this chapter, only the mass spectra generated from an EI source is discussed.

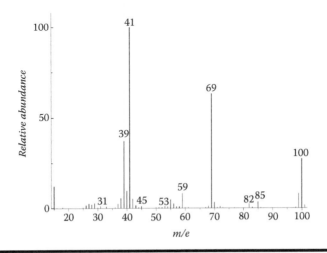

Figure 2.12   A sample mass spectrum of MMA.

1. Basic Principles of Bond Cleavage

Typically, if a bond in a compound can easily undergo cleavage, the corresponding fragment ion can be easily formed, generating a strong peak in the mass spectrum. Bond cleavage can easily occur as follows:

(1) Cleavage Occurs when the Produced Fragment Ions are Stable

For example, in the cleavage of alkylbenzene, an ion of m/e 91 is quite stable; hence, β-scission easily occurs:

$$\text{\Large ⬡}-CH_2 \underset{\raise2pt{\vdots}}{+} R \;\overset{\neg +}{} \xrightarrow{-R^{\cdot}} \text{\Large ⬡}- CH_2 \overset{\neg +}{} \longrightarrow \text{\Large ⬡}_{\oplus} \qquad m/e\ 91 \qquad (2.7)$$

For branched hydrocarbons, the order of stability is as follows:

$$\begin{array}{ccc}
R_2 & R_2 & C \\
| & | & | \\
R_1-\overset{+}{C}- > R_1-\overset{+}{C}- > R_1-\overset{+}{C}- \\
| & | & | \\
R_3 & H & H
\end{array} \qquad (2.8)$$

Hence, hydrocarbons are prone to cleavage from the branched carbon.

For compounds with atoms other than carbon, e.g., nitrogen and oxygen, β-scission can easily occur as the unshared electron pair in these atoms can stabilize carbocations:

$$\text{\Large ⬡}-CH_2-CH_2-\overset{+\cdot}{N}H_2 \xrightarrow{-\text{⬡}-\overset{\cdot}{C}H_2} CH_2=\overset{+}{N}H_2 \qquad (2.9)$$

In carbonyl-containing compounds, the unshared electron pair on the oxygen atom stabilizes the cation, facilitating α-scission:

$$\begin{array}{ccc}
R-C-R' & C-R' & {}^+C-R' \\
\diagdown\ \| & \|\|\| & \| \\
O & O & O \\
{}^{+\cdot}
\end{array} \xrightarrow{-R\cdot} \quad \longleftrightarrow \qquad (2.10)$$

A single arrow ⌣⟶ represents the transfer of one electron, and a double arrow ⌣⟹ represents the transfer of two electrons.

(2) Cleavage Occurs with the Elimination of Small Neutral Molecules

Bond cleavage easily occurs if the elimination of small neutral molecules, such as H_2O, CO, NH_3, H_2S, CH_3COOH, and CH_3OH, is possible.

$$\underset{M-17}{[M-OH]^+} \xleftarrow[\text{Difficult}]{-{}^{\cdot}OH} [ROH]^{+\cdot} \xrightarrow[\text{Easy}]{-H_2O} \underset{M-18}{[M-H_2O]^{+\cdot}} \qquad (2.11)$$

$$\underset{M-59}{[M-OAc]^+} \xleftarrow[\text{Difficult}]{} [ROAc]^+ \xrightarrow[\text{Easy}]{-AcOH} \underset{M-60}{[M-AcOH]^+} \qquad (2.12)$$

Table 2.3 Energies of Bonds (kJ/mol)

Bond	C–H	C–C	C=C	C≡C	C–N	C=N	C≡N	C–O
Energy	409.20	345.60	607.10	835.13	304.60	615.05	889.52	359.00
Bond	C=O	C–S	C=S	C–F	C–Cl	C–Br	C–I	O–H
Energy	748.94	271.96	535.55	485.34	338.90	284.51	213.38	462.75

(3) Cleavage is Favorable from an Energy Perspective

Obviously, a weak bond can be easily cleaved. Table 2.3 summarizes the energies of bonds. As compared to the C–C bond, C–N, C–S, C–Cl, C–Br, and C–I bonds are relatively weak; hence, these bonds are easily cleaved. Conversely, the C–O, C–F, O–H, and C–H are relatively strong and are not easily cleaved.

Besides, cleavage easily occurs if it can favor the formation of conjugated systems, release ring stress, decrease steric hindrance, result in the formation a stable transition state (typically, a six-membered ring), or facilitate rearrangement reactions.

2. Fragmentation Modes of Cations

Cation fragmentation mainly occurs by simple cleavage and rearrangement.

(1) Simple Cleavage

Simple cleavage occurs with the scission of only one bond and the loss of one radical. The formed fragment is part of the original structure. Some typical processes of cleavage are as follows:

a. σ-scission

Typically, σ-scission is observed in saturated alkanes. A series of EE$^+$ with a mass interval of 14 are observed in the low m/e region. Equation 2.13 shows the general formula. The formation possibilities of two cations depend on their stabilities. If branches are present, a high branching degree implies high probability for cleavage. The possibility of cleavage for branched cations follows the order of $\overset{+}{CR_3} > \overset{+}{CHR_2} > \overset{+}{CH_2R} > \overset{+}{CH_3}$.

$$R-\overset{|}{\underset{|}{C}}-\overset{|}{\underset{|}{C}}-R'^{\rceil\dot+} \longrightarrow \begin{cases} R-\overset{|}{C^+} + \cdot\overset{|}{\underset{|}{C}}-R' \\ R-\overset{|}{\underset{|}{C}}\cdot + {}^+\overset{|}{\underset{|}{C}}-R' \end{cases} \qquad (2.13)$$

b. α-scission

For compounds with atoms other than carbon, e.g., R–X (X = F, Cl, Br, I, NR$'_2$, SR', and OR'), Equations 2.14 and 2.15 show the cleavage mode. Similarly, the stability of the cation is the deciding factor for the formation of the positively charged cation fragment.

$$R-X^{\daleth\,\dot+}\left\{\begin{array}{ll} R^+ \ + \ \cdot X & (2.14) \\[2em] R\cdot \ + \ X^+ & (2.15) \end{array}\right.$$

In addition, α-scission also occurs in ketones:

$$-R'\!+\!\overset{\overset{\dot+}{O}}{\underset{\displaystyle\|}{C}}\!+\!R-\ \longrightarrow\ \left\{\begin{array}{ll} {}^+O\equiv C-R \ + \ \cdot R' & (2.16) \\[1em] {}^+O\equiv C-R' \ + \ \cdot R & (2.17) \end{array}\right.$$

R^+ and R'^+ possibly form, albeit their intensities are quite weak.

c. β-scission

If C=C is present in a molecule, allyl-type scission predominantly occurs instead of bond scission between C=C and α-C:

$$R-\overset{|}{\underset{|}{C}}=\overset{|}{C}-\overset{|}{C}-R' \ \xrightarrow{-e}\ R-C^{\pm}C-\overset{\frown}{C}\!+\!R' \ \longrightarrow\ R-\overset{|}{\underset{|}{C}}-\overset{+}{\underset{|}{C}}=\overset{|}{C} \ + \ \cdot R' \quad (2.18)$$

If atoms other than carbon are present, β-scission can also easily occur:

$$R-\overset{+}{\overset{\frown}{X}}\overset{\frown}{C}\!+\!\overset{\frown}{C}-R' \ \longrightarrow\ R-\overset{+}{X}=C\!\!\big< \ + \ \cdot\overset{|}{\underset{|}{C}}-\overset{|}{C}\!\!\big<$$
$$\underset{\alpha\ \ \ \beta}{} \qquad\qquad \downarrow \qquad\qquad (2.19)$$
$$R-\overset{..}{X}-\overset{+}{C}\!\!\big<$$

Here, X = O, N, and S.

In summary, OE$^+$ loses a radical and forms EE$^+$ by simple cleavage, which is expressed as follows:

$$OE^{+\cdot} \xrightarrow{\ \text{simple breakage}\ } EE^+ + \text{radical} \qquad (2.20)$$

(2) Rearrangement

Rearrangement starts from the transfer of a hydrogen atom, with the simultaneous cleavage of more than one bond. The structure of so-formed fragment ion cannot be identified from the original structure. Typical rearrangement modes are as follows:

a. The McLafferty rearrangement

$$(2.21)$$

If unsaturated groups and γ-H are present, the McLafferty rearrangement occurs via a six-membered-ring transition state, where R = alkyl, -H, -OH, -OR, or -NH$_2$ (Equation 2.21).

b. Reverse Diels–Alder Reaction

m/e 28   *m/e* 82   *m/e* 54

$$(2.22)$$

Conjugated diene ions are produced from the reverse Diels–Alder reaction of cycloalkene fragment ions. In Equation 2.22, the ion with an m/e 28 is more likely to be produced.

c. Loss of Neutral Molecules from Heteroatom-containing Compounds
   For example:

$$(2.23)$$

Here, X = OCOCH$_3$, OH, SH, F, Cl, Br, and I. When n = 0–4, neutral molecules are lost, e.g., the dehydration of alcohols.

In summary, rearrangement reactions often occur followed by simple cleavage. The resultant fragment ions are different from the original structure. If the parent ion is EE$^+$, the daughter ion is EE$^+$; if the parent ion is OE$^{+\cdot}$, the daughter ion is OE$^{+\cdot}$. Their general equations are as follows:

$$\text{EE}^+ \xrightarrow[\text{Rearrangement}]{\text{H}} \xrightarrow{\text{Simple cleavage}} \text{EE}^+ + \text{Neutral molecule} \qquad (2.24)$$

$$\text{OE}^{\pm} \xrightarrow[\text{Rearrangement}]{\gamma\text{H}} \xrightarrow{\text{Simple cleavage}} \text{OE}^{\pm} + \text{Neutral molecule} \qquad (2.25)$$

Other fragmentation modes of cations, in addition to simple cleavage and rearrangement (e.g., complex cleavage and double rearrangement), are also observed. Besides, cleavage and rearrangement modes are not limited to the above cases. For brevity, these complex cases are not introduced here.

2.3.3.3 Mass Spectra of Typical Compounds

Various compounds exhibit specific characteristics in their mass spectra, corresponding to different structures and fragmentation modes. Hence, it is imperative to obtain information regarding the mass spectra of typical compounds for the spectrometric analysis of unknown compounds.

1. Alkanes

    Typically, σ-scission occurs in alkanes, producing a series of EE$^+$ with a mass interval of 14 (-CH$_2$). The relative abundance of the molecular ion decreases with increasing molecular weight and branching degree. The highest peak is often observed at C$_3$–C$_6$ (m/e 43, 57, 71). Figure 2.13a and b shows examples.

2. Ethers

    β-scission predominantly occurs in ethers, with the formation of a series of EE$^+$ (m/e 45, 59, 73, 87...).

$$R \!+\! CH_2\!-\!\overset{+\cdot}{O}\!-\!CH_2 \!+\! R' \xrightarrow{\ -R'\cdot\ } R\!-\!CH_2\!-\!\overset{+}{O}\!=\!CH_2$$
$$\xrightarrow[\ -R\cdot\ ]{} CH_2\!=\!\overset{+}{O}\!-\!CH_2\!-\!R' \tag{2.26}$$

In addition, α-scission possibly occurs, and R$^+$ or OR'$^+$ is produced from the α-scission of R-OR'. R$^+$ has better stability than OR'$^+$. Figure 2.14 shows the mass spectrum of isopropyl ether, where the intensity of the m/e 43 peak is greater than that of the m/e 59 peak.

The base peak of m/e 45 is obtained via the fragmentation of m/e 87:

$$\begin{array}{c} H_3C\!\!\diagdown \\ \diagup \\ H_3C \end{array}\!\!CH\!-\!O\!-\!CH\!\!\begin{array}{c}\diagup CH_3 \\ \\ \diagdown CH_3\end{array} \xrightarrow[-CH_3]{-e} \begin{array}{c} H\!-\!CH_2 \\ \\ CH_3\!-\!CH\!=\!\overset{+}{O}\!-\!CH\!-\!CH_3 \\ m/e\ 87 \end{array}$$
$$\xrightarrow[\ -CH_2=CH-CH_3\ ]{} CH_3\!-\!CH\!=\!\overset{+}{O}H \tag{2.27}$$
$$m/e\ 45$$

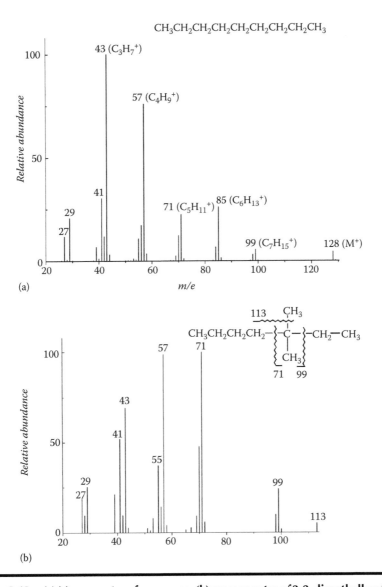

Figure 2.13 (a) Mass spectra of n-nonane; (b) mass spectra of 3,3-dimethylheptane.

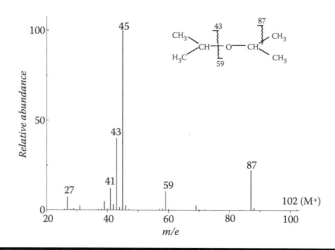

Figure 2.14 Mass spectrum of isopropyl ether.

3. Alcohols

As shown in Equation 2.11, alcohols are not stable and easily eliminate H_2O. Hence, molecular ion peaks observed for primary alcohols and secondary alcohols are quite weak, while those of tertiary alcohols are not detected. β-scission can easily occur, affording oxonium ions:

$$R-\overset{\overset{+\cdot}{O}H}{\underset{\underset{H}{|}}{C}}-R' \longrightarrow R-\overset{\overset{+}{O}H}{\overset{\|}{C}H} + \cdot R' \tag{2.28}$$

Long-chain aliphatic alcohols easily eliminate H_2O and alkene simultaneously, affording $OE^{+\cdot}$ with m/e 42, 56, 70, 84, 98...:

$$\begin{array}{c} \xrightarrow[-C_2H_4]{-H_2O} H_2C-CH-R^{\overline{}+} \end{array} \tag{2.29}$$

4. Ketones

α-scission easily occurs in ketones following the routes shown in Equations 2.16 and 2.17, with the formation of m/e 57 and m/e 85 (Figure 2.15).

If γ-H is present, rearrangement occurs, affording an enol ion (Equation 2.21). If different substituents (X) are present, different mass charge ratios are observed (Table 2.4). Hence, the substituent can be identified by m/e. As shown in Figure 2.15, an m/e 72 corresponds to an ethyl substituent.

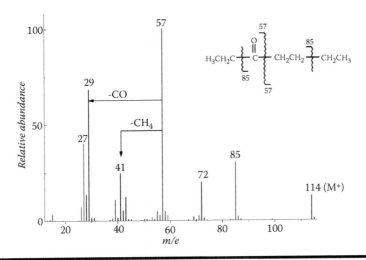

Figure 2.15   Mass spectrum of 3-heptanone.

Table 2.4   Mass-to-Charge Ratio of Rearrangement Ions and Corresponding Substituents in Equation 2.21

X	H	CH$_3$	NH$_2$	OH	C$_2$H$_5$	OCH$_3$
m/e	44	58	59	60	72	74

5. Esters

Typically, carboxylates form obvious molecular ions. R-C≡O$^+$ and $^+$O≡C-OR′ ions are easily formed via α-scission. For example, methyl esters are cleaved into the following ions:

$$
\begin{array}{cccc}
\overset{+}{\underset{\overset{|||}{R-C}}{O}} & \overset{+}{\underset{\overset{|||}{C-OCH_3}}{O}} & \overset{+}{O-CH_3} & R^+ \\
M-31 & m/e\,59 & m/e31 & M-59
\end{array}
$$

Figure 2.12 shows the mass spectrum of MMA, with fragment ions observed at m/e 31, 59, 69, and 41.

Acetic acid can be easily eliminated from acetate, affording an ion with m/e 60 (Equation 2.12).

6. Aromatics

Typically, aromatic compounds exhibit strong molecular ion peaks. Characteristic fragments of aromatic rings are observed at m/e 39, 51, 65, and 77. Figure 2.16 shows the mass spectrum of toluene. The base peak (m/e 91) is C$_7$H$_7^+$, which can be further cleaved, affording C$_5$H$_5^+$ (m/e 65) and C$_3$H$_3^+$ (m/e 39).

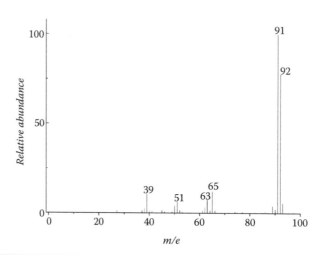

Figure 2.16   Mass spectrum of toluene.

For a substituted benzene ring, the fragmentation modes are ranked from easy to difficult as follows:

$$C_6H_5 - \overset{O}{\overset{||}{C}} - CH_3]^{+\cdot} \longrightarrow C_6H_5 - \overset{O^+}{\overset{|||}{C}} + {\cdot}CH_3$$

$$C_6H_5 - \underset{\underset{CH_3}{|}}{\overset{\overset{CH_3}{|}}{C}} - CH_3]^{+\cdot} \longrightarrow C_6H_5 - \underset{\underset{CH_3}{|}}{\overset{\overset{CH_3}{|}}{C}{+}} \; + \; {\cdot}CH_3$$

$$C_6H_5 - \overset{O}{\overset{||}{C}} - OCH_3]^{+\cdot} \longrightarrow C_6H_5 - \overset{O^+}{\overset{|||}{C}} + {\cdot}OCH_3$$

$$C_6H_5 - N \overset{CH_3}{\underset{CH_3}{\diagup}}]^{+\cdot} \longrightarrow C_6H_5 - \overset{+}{N} \overset{CH_3}{\underset{CH_2}{\diagup}} \; + \; {\cdot}H$$

$$C_6H_5 - C \overset{H}{\underset{O}{\diagup}}]^{+\cdot} \longrightarrow C_6H_5 - C \equiv O^+ + {\cdot}H$$

$$C_6H_5-O-CH_3 \quad]^{+} \diagdown \begin{array}{l} \nearrow C_6H_5O^+ + \ ^\cdot CH_3 \\ \searrow C_6H_6 \ ]^{+\cdot} + CH_2O \end{array}$$

$$C_6H_5-I \ ]^{+\cdot} \longrightarrow C_6H_5^+ + \ ^\cdot I$$

$$C_6H_5-OH \ ]^{+\cdot} \longrightarrow C_5H_6 \ ]^{+\cdot} + CO$$

$$C_6H_5-CH_3 \ ]^{+\cdot} \longrightarrow C_6H_5-CH_2^+ + \ ^\cdot H$$

$$C_6H_5-Br \ ]^{+\cdot} \longrightarrow C_6H_5^+ + \ ^\cdot Br$$

$$C_6H_5-NO_2 \quad]^{+} \diagdown \begin{array}{l} \nearrow C_6H_5^+ + \ ^\cdot NO_2 \\ \searrow C_6H_5O^+ + \ ^\cdot NO \end{array}$$

$$C_6H_5-NH_3 \ ]^{+\cdot} \longrightarrow C_5H_6 \ ]^{+\cdot} + HCN$$

$$C_6H_5-Cl \ ]^{+\cdot} \longrightarrow C_6H_5^+ + \ ^\cdot Cl$$

$$C_6H_5-CN \ ]^{+\cdot} \longrightarrow C_6H_4 \ ]^{+\cdot} + HCN$$

$$C_6H_5-F \ ]^{+\cdot} \longrightarrow C_6H_5^+ + \ ^\cdot F$$

2.3.3.4 Spectral Interpretation of Unknown Compounds

Although characteristic fragmentation modes are known, it is quite difficult to analyze a mass spectrum. A majority of manufacturers of mass spectrometers provide standard libraries for searching, but the result is only based on a similarity calculation and is possibly not reliable. Furthermore, if there is no corresponding compound in the library, a reasonable result cannot be obtained. Hence, it is necessary to understand the basics of spectral interpretation. It includes the following three steps:

1. Determination of the Molecular Ion and Formula
   First, the molecular ion peak should be determined according to the requirements in Section 2.3.1, and a formula should be proposed on the basis

of m/e. For this purpose, the Table of Isotope Abundance Values by Beynon [6] is helpful.

Then, the degree of unsaturation (Ω) should be calculated according to Equation 2.30, with the number of carbon, hydrogen, and nitrogen atoms (oxygen and sulfur atoms are not considered):

$$\Omega = 1 + n_C + \frac{n_N - n_H}{2} \tag{2.30}$$

If $\Omega = 0$, the molecule is saturated; if $\Omega = 4$, a benzene ring is possibly present in the molecule.

2. Search of Characteristic Fragment Ions

Strong and important peaks in a mass spectrum should be labeled as $OE^{+\cdot}$ and EE^+. Often, more EE^+ peaks are formed by simple cleavage. If $OE^{+\cdot}$ is observed in the high m/e region, the possibility of rearrangement reactions must be considered.

First, the elimination of neutral fragments should be ascertained from the high m/e peaks. This loss often provides information about the substituents. For example, M-15 possibly represents the loss of $-CH_3$; hence, a methyl group is possibly present in the molecule. M-18 possibly represents the loss of H_2O; hence, alcohols are possibly present. Table 2.5 summarizes the common source of neutral fragments from high m/e peaks.

Second, a series of EE^+ in the low m/e region should also be determined. For example, the fragments of m/e 15, 29, 43, 57, 71, 85, and 99...correspond to alkane cations CnH^+_{2n+1}, while alkyl amine $(-CH_{2n}NH_2)$ produces an EE^+ series with m/e 30, 44, 58, and 72... Table 2.6 summarizes the common source of low m/e fragments.

3. Proposal of Possible Structural Formula and Verification

It is difficult to identify every peak in a mass spectrum, but it is possible and essential to identify characteristic peaks, especially strong peaks with high m/e.

Let us practice using three simple examples.

Example 2.1 explains structural determination from isotopic peaks. Figure 2.17 shows the mass spectrum of a common organic solvent.

The first task involves finding the molecular ion. If peak with the highest m/e 88 is believed to be a candidate, M-2 and M-4 peaks are considered. The abundance of the M-4 peak (m/e 84) is quite high. It is not possible that the M-4 peak is significantly greater than the molecular ion peak. Hence, peaks with m/e 84, 86, and 88 possibly correspond to a group of isotope ions; among these peaks, the m/e 84 peak corresponds to the molecular ion. Hence, the relative abundance ratio of these peaks is M:M+2:M+4 = 9:6:1. This abundance ratio implies that two chlorine atoms are possibly present in the molecule. From a molecular weight of 84, the proposed molecular

Table 2.5   Source of Neutral Fragments from High m/e Peaks

M–X	*X*	*Common Source*
M-1	H	Alkyl nitrile, aldehyde, amine, low-molecular-weight fluoroalkane
M-15, 29, 43, 57…	C_nH_{2n+1}	Loss of alkyl groups by α-scission
M-16	O	Sulfone and nitroxide
	NH_2	Aromatic amine (weak)
M-17	OH	Acid and oxime
M-18, 32, 46, and 60…	$H_2O+C_nH_{2n}$	Alcohol (especially primary), high-molecular-weight aldehyde, ketone, ester, ether, and o-methyl aromatic acid
M-19, 33, and 47…	$C_nH_{2n}F$	Fluoroalkane
M-20	HF	Fluoroalkane
M-26, 40, 54, and 68…	$C_2H_2(CH_2)_n$	Aromatic hydrocarbon
M-27	HCN	Heterocycles with nitrogen, aromatic amine, and oxide
M-26, 40	$C_nH_{2n}CN$	R-CN, when R⁺ is very stable
M-28	N_2	ArN=NAr
	CO	Aromatic ketone, phenol, cyclic ketone, and quinine
M-28, 42, 56, and 70…	C_nH_{2n}	The McLafferty rearrangement and similar rearrangement reactions
M-29, 43, 57, and 71…	$C_nH_{2n+1}CO$	$C_nH_{2n+1}COR$ (R⁺ is very stable)
M-46	CH_2O_2	Aliphatic diacid
	$CH_2\begin{smallmatrix}O-\\O-\end{smallmatrix}$	
M-48	SO	Aromatic sulfoxide
M-59 and 73…	$C_nH_{2n+1}COO$	ROCOR' or RCOOR', when R⁺ is very stable, and R' is small

(Continued)

Table 2.5 (Continued)   Source of Neutral Fragments from High m/e Peaks

M–X	X	Common Source
M-60 and 74	$C_nH_{2n+1}COOH$	R'COOR, when R+ is very stable, and R' is small
M-60	COS	Sulfocarbonate
M-64	SO_2	RSO_2R' (sulfone) and $ArSO_2OR$ (aromatic sulfonic ester)
M-79	Br	RBr
M-127	I	RI

Table 2.6   Common Source of Low m/e Fragment Ions

m/e	General Formula	Common Source
15, 29, 43, 57, 71, and 85...	C_nH_{2n+1} $C_nH_{2n+1}CO$	Alkyl Carbonyl, cycloalcohol, and cycloether
19, 33, 47, 61, 75, and 89...	$C_nH_{2n+1}O_2$	Ester, acetal, and hemiacetal
26, 40, 54, 68, 82, 96, and 110...	$C_nH_{2n}CN$	Alkyl nitrile and dicyclomine
30, 44, 58, 72, and 86...	$C_nH_{2n+2}N$ $C_nH_{2n+2}NCO$	Aliphatic amine Amide, urea, and urethane
31, 45, 59, and 73...	$C_nH_{2n+1}O$ $C_nH_{2n-1}O_2$ $C_nH_{2n+3}Si$ $C_nH_{2n-1}S$	Aliphatic alcohol and ether Acid, ester, cycloacetal, and ketal Alkyl silane Thia-cycloalkane and unsaturated and substituted sulfide
31, 50, 69, 100, and 119, 131, 169, 181, and 193...	C_nF_m	Perfluoroalkane and perfluorokerosene (PFK)
33, 47, 61, 75, and 89...	$C_nH_{2n+1}S$	Mercaptan and sulfide

(Continued)

Table 2.6 (Continued)   Common Source of Low m/e Fragment Ions

m/e	General Formula	Common Source
38 (39, 50, 51, 63, 64, 75, and 76) 39 (40, 51, 52, 65, 66, 77, and 78)		Aromatics with electron-withdrawing groups Aromatics with electron-donating groups
39, 53, 67, 81, 95, and 109...	C_nH_{2n-3}	Diene, alkyne, and cycloalkene
41, 55, 69, 83, and 97...	C_nH_{2n-1}	Alkene and cycloalkane
55, 69, 83, 97, and 111...	$C_nH_{2n-1}CO$ $C_nH_{2n}N$	Cycloketone, cycloalcohol, and ether Vinyl amine and cycloamine
56, 70, 84, and 98...	$C_nH_{2n}NCO$ $C_nH_{2n+2}NO$	Alkyl isocyanate Amide
60, 74, 88, 102...	$C_nH_{2n}NO_2$ NO_2	Nitrite
46	$C_nH_{2n+1}O_3$	Carbonate
63, 77, and 91... (45, 57, 58, 59, 69, 70, 71, and 85)		Thiophene
69, 81–84, 95–97, 107–110		Compounds with sulfur connected to the aromatic ring
72, 86, 100...	$C_nH_{2n}NCSO$	Alkyl isorhodanate
77, 91, 105, and 119...	$C_6H_5C_nH_{2n}$	Alkyl benzene
78, 92, 106, and 120...	$C_5H_4NC_nH_{2n}$	Pyridine derivatives and aromatic amine
79, 93, 107, and 121...	C_nH_{2n-5}	Terpene and derivatives
81, 95, and 109...	$C_nH_{2n-1}O$	Alkyl furan, cycloalcohol, and ether
83, 97, 111, and 125...	$C_4H_3SC_nH_{2n}$	Alkyl thiophene
105, 119, and 133...	$C_nH_{2n+1}C_6H_4CO$	Alkylbenzoyl compound

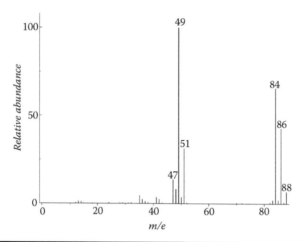

Figure 2.17   Mass spectrum of an organic solvent.

formula is CH_2Cl_2, i.e., dichloromethane. Finally, this result needs to be verified from the spectrum:

$$CH_2Cl_2^{\;\cdot+} \xrightarrow{\;-\;\cdot Cl\;} CH_2Cl^+$$
$$m/e\;49,51$$

The base peak is formed by the loss of a chlorine radical from the molecular ion, and the relative abundance ratio of m/e 49 to m/e 51 is approximately 3:1, which conforms to the ion structure.

Example 2.2. A compound does not contain chlorine, and Figure 2.18 shows its mass spectrum.

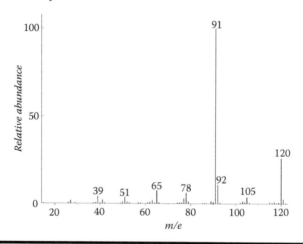

Figure 2.18   Mass spectrum of an unknown compound.

The peak with the highest mass (m/e 120) meets the criterion of molecular ions and can be assumed to be the molecular ion peak. There is no characteristic isotopic peak; hence, the molecular structure needs to be deduced from the fragment ions. First, OE⁺· and EE⁺ are labeled. The peak of m/e 92 corresponds to OE⁺·, while other peaks correspond to EE⁺. The EE⁺ series of m/e 39, 51, 65, 77, and 91 are observed, indicating the presence of a benzyl ion. The peaks of m/e 105 and 91 correspond to the M-15 and M-29 peaks, indicating the presence of methyl and ethyl groups, respectively. The OE⁺· peak is possibly related to rearrangement; hence, γ-H is possibly present in the compound. From the above conclusions, the compound is possibly n-propylbenzene. The characteristic ions and their formation are verified as follows:

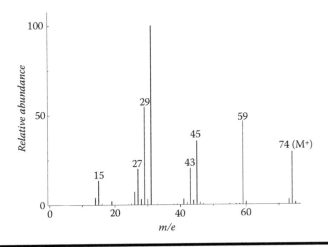

Hence, the unknown compound is n-propylbenzene.

Example 2.3 The molecular formula of an unknown compound is $C_4H_{10}O$, and Figure 2.19 shows the mass spectrum.

From the molecular formula, the peak of m/e 74 is the molecular ion peak. The degree of unsaturation is calculated to be 0, indicating that the

Figure 2.19 Mass spectrum of an unknown compound.

compound does not contain a double bond and possibly corresponds to aliphatic alcohols or ethers. The peak of m/e 59 corresponds to the M-15 peak, indicative of -CH$_3$. From Table 2.6, the EE$^+$ series of m/e 31, 45, and 59 possibly correspond to C$_n$H$_{2n+1}$O or C$_n$H$_{2n-1}$O$_2$, but the latter does not coincide with the molecular formula, and hence not possible. The possible structural formulas are as follows:

(a) C$_2$H$_5$–O–C$_2$H$_5$
(b) CH$_3$–O–CH$_2$–CH$_2$–CH$_3$
(c) HO–CH$_2$–CH$_2$–CH$_2$–CH$_3$

If the structural formula is (c), an M-18 peak must be present. As an M-18 peak is absent in the spectrum, compound (c) can be excluded. The M-15 peak in compound (b) must not be strong; hence, the most possible structure is compound (a). Let us verify further:

$$C_2H_5-\overset{+\cdot}{\ddot{O}}-C_2H_5 \xrightarrow{-CH_3} C_2H_5-\overset{+}{\ddot{O}}=CH_2 \xrightarrow[\text{rearrangement}]{-C_2H_4} \overset{+}{H\ddot{O}}=CH_2$$
$$\phantom{C_2H_5-\overset{+\cdot}{\ddot{O}}-C_2H_5 \xrightarrow{-CH_3}} \text{m/e 59} \qquad\qquad \text{m/e 31}$$

$$C_2H_5-\overset{+\cdot}{\ddot{O}}-C_2H_5 \xrightarrow{-H\cdot} C_2H_5-\overset{+}{\ddot{O}}=CHCH_3 \xrightarrow[\text{rearrangement}]{-C_2H_4} \overset{+}{H\ddot{O}}=CHCH_3$$
$$\phantom{C_2H_5-\overset{+\cdot}{\ddot{O}}-C_2H_5 \xrightarrow{-H\cdot} C_2H_5-\overset{+}{\ddot{O}}=CHCH_3} \text{m/e 45}$$

In summary, although MS is a powerful qualitative method, obtaining the structural identification of an unknown compound only by MS is difficult, and other analytical methods are required.

2.4 Applications of PGC in Polymer Materials

2.4.1 Qualitative Identification

Various polymers exhibit their own pyrolysis modes at specific temperatures and produce characteristic pyrograms, which can be utilized for the qualitative

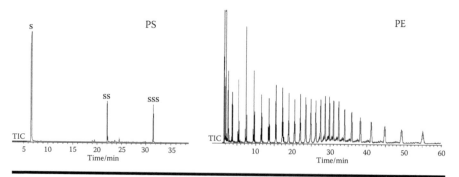

Figure 2.20 Pyrograms of PS and PE.

identification of polymers. Figure 2.20 shows the pyrograms of PS and PE as examples. The main pyrolysates obtained from PS are styrene (S), dimer (SS), and trimer (SSS). The pyrolysates of PE include a series of n-alkenes, which are easily differentiated simply from the "shape" of the pyrograms.

If the "shape" of the pyrograms of polymers is not typical and easily distinguishable, the structure of the main pyrolysates of an unknown polymer can be deduced via the analysis of their mass spectra, and the polymer structure may be proposed. Figure 2.21a shows the pyrogram of an unknown copolymer. The mass spectra of the three characteristic peaks labeled in Figure 2.21a are shown in Figure 2.21b, c, and d, respectively. By searching the NIST standard library and spectra analysis, numbers 1, 2, and 3 correspond to butadiene, methyl methacrylate (MMA), and styrene, respectively. Hence, the unknown copolymer is methyl methacrylate–butadiene–styrene terpolymer (MBS).

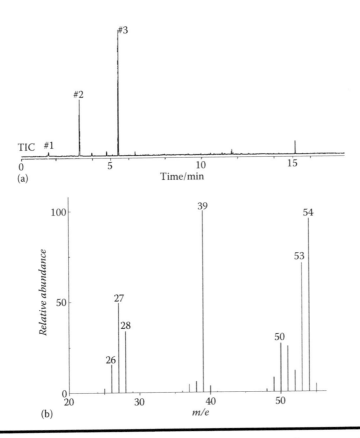

Figure 2.21 **Pyrogram of an unknown copolymer (a) [2] and mass spectra with three characteristic peaks: (b) 1, (c) 2, and (d) 3.** (*Continued*)

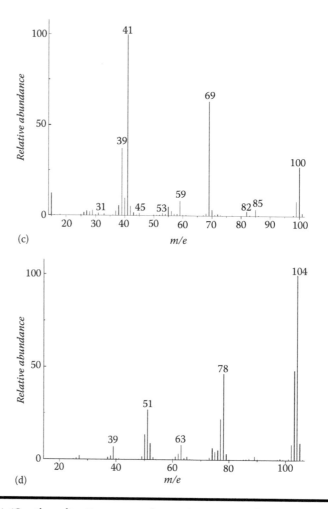

Figure 2.21 (Continued) **Pyrogram of an unknown copolymer (a) [2] and mass spectra with three characteristic peaks: (b) 1, (c) 2, and (d) 3.**

As PGC is sensitive to the number of carbons, and homologs with different numbers of carbons can be clearly separated in the pyrograms, PGC is very advantageous for the identification of similar polymers in the same group. For example, the IR spectra of various nylons are quite similar, caused by the presence of the same functional groups. Nevertheless, obvious differences in their pyrograms are observed (Figure 2.22). Characteristic pyrolysates are different because of their different pyrolysis routes. Let us take nylon-6 and nylon-6,6 as examples. The

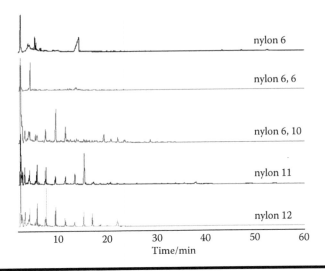

Figure 2.22 Pyrograms of various nylons.

characteristic pyrolysate of nylon-6 is the monomer caprolactam, while that of nylon-6,6 is cyclopentanone.

The copolymerization of terephthalic acid with various diols having different numbers of carbon atoms affords a series of aromatic polyesters, which are clearly differentiated by PGC. Figures 2.23 and 2.24 show the pyrolysis routes of aromatic polyesters and their pyrograms at $n = 2$–6 [7]. At different n, the number of carbons associated with c, e, f, g, and h is different, resulting in different retention times in their pyrograms.

2.4.2 Differentiation of Blends and Copolymers

Blends and copolymers with the same monomers exhibit the same chemical composition as well as similar structures, with the main difference associated with the conjunction of monomer units. In blends of homopolymers, only covalent

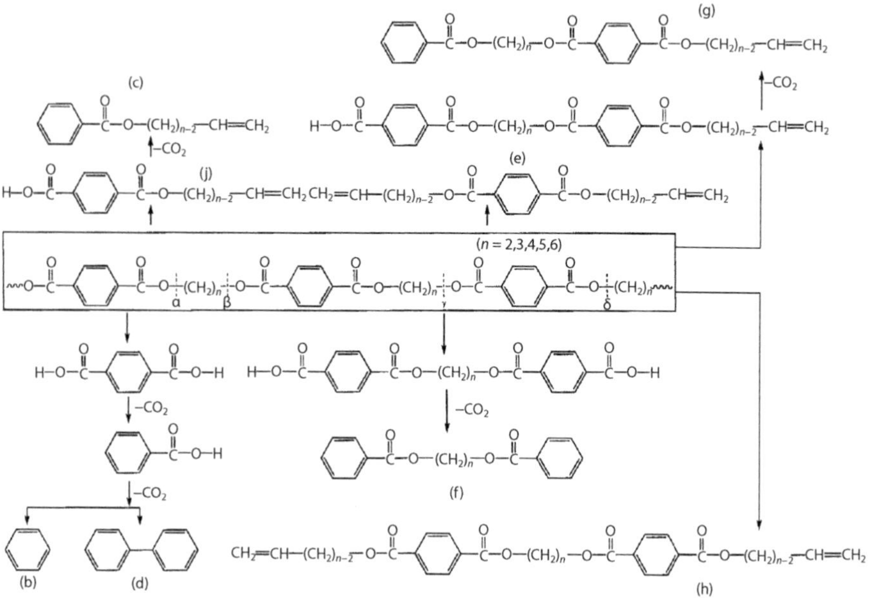

Figure 2.23 Pyrolysis routes of aromatic polyesters. (From Ohtani, H. et al., *Analytical Science*, **1986, 2(2), 179–182.)**

conjunctions between the same monomer units are present. Among homopolymers, only intermolecular interactions, e.g., Van der Waals forces, are present. On the other hand, in copolymers, not only covalent conjunctions between the same monomer units, but also covalent conjunctions between different monomer units, exist. This subtle difference can be identified by PGC. Figure 2.25 shows the pyrograms of polymethyl acrylate (PMA), PS, and methyl acrylate–styrene copolymer (PMA–S) [8]. In the pyrograms of PMA and PS, MA and S monomers and their oligomers are observed. In the pyrogram of PMA–S, heterodimers and heterotrimers, as well as monomers, are observed. These peaks are not present in the pyrogram of the PMA/PS blend, which should serve as an additional result of the pyrograms of the two homopolymers in the absence of secondary reactions.

2.4.3 *Sequence Distribution of Copolymers*

As mentioned above, hetero-oligomers are produced during the pyrolysis of copolymers. Their yields (concentrations) depend not only on the composition but also

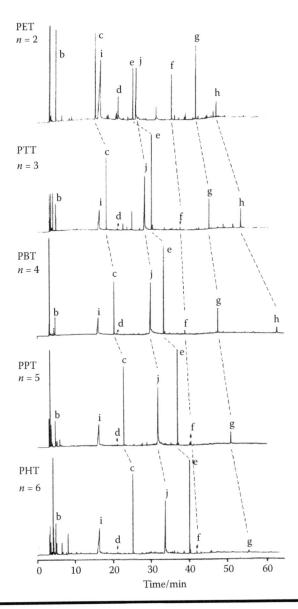

Figure 2.24 Pyrograms of aromatic polyesters at 590°C. (From Ohtani, H. et al., *Analytical Science*, **1986, 2(2), 179–182.)**

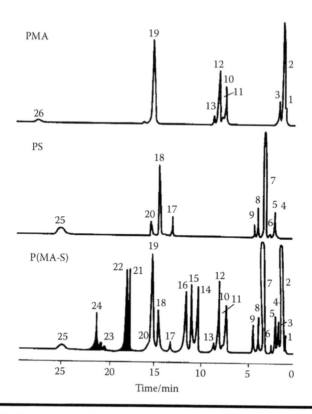

Figure 2.25 Pyrograms of PMA, PS, and P(MA-S). (From Tsuge, S. et al., *Macromolecules*, 1975, 8(6), 721–725.)

on the sequence distribution of monomers. Let us take an A–B copolymer as an example. The number average sequence length (NASL) $n(A)$ [9] can be defined as follows:

$$n(A) = \frac{\text{Total number of } A}{\text{Total blocks of } A} \tag{2.31}$$

With different connections, four A-centered triads can produce A:

$$\text{\textasciitilde A-A-A\textasciitilde} \xrightarrow{K_1} A \tag{2.32}$$

$$\text{\textasciitilde A-A-B\textasciitilde} \xrightarrow{K_2} A \tag{2.33}$$

$$\text{\small\textasciitilde\textasciitilde\textasciitilde} B\text{-}A\text{-}A \text{\small\textasciitilde\textasciitilde\textasciitilde} \xrightarrow{K_3} A \qquad\qquad (2.34)$$

$$\text{\small\textasciitilde\textasciitilde\textasciitilde} B\text{-}A\text{-}B \text{\small\textasciitilde\textasciitilde\textasciitilde} \xrightarrow{K_4} A \qquad\qquad (2.35)$$

Here, K_1, K_2, K_3, and K_4 represent the possibilities of producing A from the four triads, respectively. In most cases, $K_2 = K_3$. The higher the possibility, the higher the yield of A. As the yield of A depends also on the concentration of triads with A in a copolymer (AAA+AAB+BAA+BAB), it can also be utilized to investigate the sequence distribution of copolymers.

Similarly, diads and triads can also be utilized to study the sequence distribution of copolymers. For example, NASL can be expressed by triads as follows:

$$n(A) = \frac{N_{AAA} + N_{AAB+BAA} + N_{BAB}}{1/2 N_{BAA+AAB} + N_{BAB}} \qquad\qquad (2.36)$$

$$n(B) = \frac{N_{BBB} + N_{ABB+BBA} + N_{ABA}}{1/2 N_{ABB+BBA} + N_{ABA}} \qquad\qquad (2.37)$$

Here, N_{AAA} is the number of the AAA triads. Other parameters are similarly defined.

The sequence distribution of a copolymer is calculated in three steps: (1) Identification of the six characteristic peaks from the pyrogram of a copolymer, i.e., AAA, AAB/BAA, ABA, BAB, BBA/ABB, and BBB. (2) Determination of the pyrogram of the same copolymer with a known composition (random copolymer is the best because all six peaks are observed) under the same condition as the copolymer and calculation of correction factors k_n (k_n = number of triad/peak intensity of triad) of the six peaks. (3) The calculation of the number of the six triads in the unknown copolymer using k_n and $n(A)$ and $n(B)$ according to Equations 2.36 and 2.37.

Further, the molar concentrations of A and B can be calculated from NASL:

$$A(mol\%) = \frac{n(A)}{n(A) + n(B)} \times 100\% \qquad\qquad (2.38)$$

$$B(mol\%) = \frac{n(B)}{n(A) + n(B)} \times 100\% \qquad\qquad (2.39)$$

2.4.4 Conjunction Mode

Typically, monomer units are connected to each other in a "head-to-tail" manner. If there is "head-to-head" or "tail-to-tail" structure, the stability of the polymer decreases. For example, during the pyrolysis of a vinyl chloride (V)–vinylidene chloride (D) copolymer, HCl is eliminated first, affording conjugated double bonds in the main chain. For PVC, the main pyrolysate is benzene. For polyvinylidene chloride (PVDC), the main pyrolysate is 1,3,5-trichlorobenzene. In a copolymer, if only the "head-to-tail" structure is present, the pyrolysates from different triads are benzene, chlorobenzene, 1,3-dichlorobenzene, and 1,3,5-trichlorobenzene [10].

$$V\text{-}V\text{-}V \longrightarrow \bigcirc$$

$$V\text{-}V\text{-}D, V\text{-}D\text{-}V, D\text{-}V\text{-}V \longrightarrow \bigcirc\text{-}Cl$$

$$D\text{-}D\text{-}V, D\text{-}V\text{-}D, V\text{-}D\text{-}D \longrightarrow \underset{Cl}{\bigcirc}\text{-}Cl$$

$$D\text{-}D\text{-}D \longrightarrow Cl\text{-}\underset{Cl}{\overset{Cl}{\bigcirc}}$$

If "head-to-head" or "tail-to-tail" structures are present, some new pyrolysates from the triads, such as o-dichlorobenzene, p-dichlorobenzene, and 1,2,5-trichlorobenzene, are observed. Hence, the conjunction structure of copolymers is investigated from the pyrolysates. Figure 2.26 shows pyrograms of the vinyl chloride–vinylidene chloride copolymer with different compositions. Only benzene is produced from PVC. With the introduction of D monomer units, the intensity of benzene decreases, and new peaks corresponding to vinylidene chloride are observed, as well as those corresponding to chlorobenzene and m-dichlorobenzene. With the increase of D monomer units, the intensity of chlorobenzene decreases, and the intensities of vinylidene chloride and 1,3,5-trichlorobenzene increase, until finally only vinylidene chloride and 1,3,5-trichlorobenzene can be observed in the pyrogram of PVDC. As there is no evidence for the production of o-, p-dichlorobenzene, and 1,2,5-trichlorobenzene, the concentration of the "head-to-head" or "tail-to-tail" structure can be neglected.

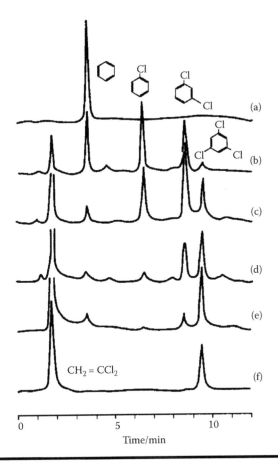

Figure 2.26 Pyrograms of vinyl chloride–vinylidene chloride copolymers. (a) PVC; (b–e) copolymers with a D content of (b) 0.127, (c) 0.281, (d) 0.598, and (e) 0.784; (f) PVDC. (From Wang, F.C.-Y., and P.B. Smith, *Analytical Chemistry.* 1996, 68(3), 425–430.)

2.4.5 Branching Degree

Branching structures, such as branch length and branching density, significantly affect the mechanical properties of PE. The branching degree is defined as the branch content per 1000 carbon atoms. Low-density polyethylene (LDPE) exhibits a relatively high branching density, and hence, good ductility but low crystallinity. LDPE is often used for preparing films. In contrast, high-density polyethylene (HDPE) exhibits much less branches, and thus good strength and high crystallinity.

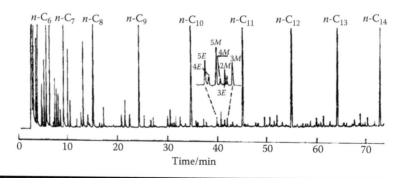

Figure 2.27 **Pyrogram of LDPE at 650°C. (2M: 2-methyldecane; 3M: 3-methyldecane; 4M: 4-methyldecane; 5M: 5-methyldecane; 3E: 3-ethylnonane; 4E: 4-ethylnonane; and 5E: 5-ethylnonane.) (From Ohtani, H. et al.,** *Macromolecules,* **17, 1984, 2557–2561.)**

For investigating the branching degree of LDPE, hydrogenation is often employed to change alkenes and dienes to alkanes; hence, branched alkenes and dienes are converted to the resultant branched alkanes (Figure 2.27) [11]. Here, a group of branched undecane peaks between n-C_{10} and n-C_{11} peaks are selected for calculation of the branching degree.

A series of ethylene-α-olefin model copolymers with known compositions are used as standards for quantitative analysis. Their branch type and content are different (Table 2.7). Figure 2.28 shows the partial pyrograms of these standards. The most important thing is to solve the correlation factor between intensities of characteristic peaks of certain branch types and their content in all standards. For example, in an ethylene–propylene copolymer (EP), only methyl branches are present in the main chain with a branching degree of 20/1000C. Characteristic peaks such as 2M, 3M, 4M, and 5M are observed in the pyrogram. Hence, the correlation factors of these peaks are calculated by dividing the branching degree by the peak intensity. For other copolymers, the correlation factors of characteristic peaks can be similarly calculated. Hence, in LDPE, the intensity of characteristic peaks is regarded as the total contribution of various branch types, and a set of equations are constructed to calculate the contents of various branches and the total branching degree.

2.4.6 *Stereoregularity*

For the industrial preparation of polyolefins, polymers with different stereoregularities, and thus different performances, are prepared by changing catalysts and conditions. Figure 2.29 shows the two steric configurations of PP. Isotactic PP with all *m* configuration or syndiotactic PP with all *r* configuration are strong and tough plastics, while atactic PP with an irregular distribution of *m* and r configurations is a viscous semisolid with no shape at room temperature.

Table 2.7 Branch Types and Content in Model Ethylene-α-olefin Copolymers

Sample	α-olefin	Branch Type	Branching Degree[a]	Sample	α-olefin	Branch Type	Branching Degree[a]
EP	Propene	Methyl	20	EHP	1-Heptene	Pentyl	12
EB	1-Butene	Ethyl	24	EO	1-Octene	Hexyl	20
EHX	1-Hexene	Butyl	18				

Note: EP: ethylene–propylene copolymer; EB: ethylene–butene copolymer; EHX: ethylene–hexene copolymer; EHP: ethylene–heptene copolymer; EO: ethylene–octene copolymer.

[a] The branching degree is determined by IR.

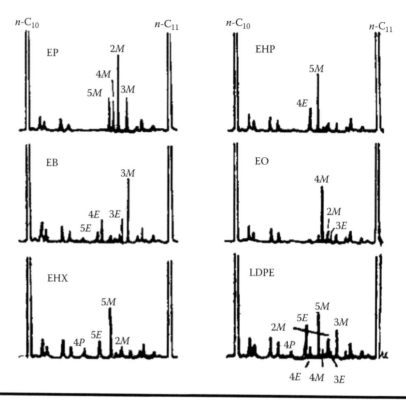

Figure 2.28 Partial pyrograms of ethylene-α-olefin model copolymers and LDPE. (From Ohtani, H. et al., *Macromolecules*, 17, 1984, 2557–2561.)

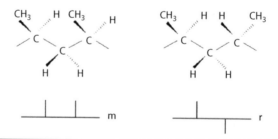

Figure 2.29 Schematic of two steric configurations of PP.

In pyrograms of these PPs, obvious differences in triads and pentads are observed. Figure 2.30 shows the pyrograms of isotactic PP and atactic PP. In pentads (C_{15}, 2,4,6,8-tetramethyl-1-undecene), the relative intensities of the three peaks are different: in isotactic PP, the left one is the highest, followed by the right one. The middle one is quite low; in atactic PP, the three peaks exhibit similar intensities. Hence, the relative intensity of pentads can be utilized to investigate the stereoregularity of PP.

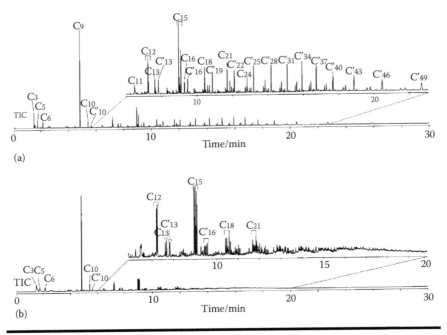

Figure 2.30 Pyrograms of (a) isotactic PP and (b) atactic PP. (Reprinted from *Pyrolysis-GC/MS Data Book of Synthetic Polymers*, Tsuge, S., H. Ohtani, and C. Watanabe, Copyright 2011, with permission from Elsevier.)

2.4.7 Volatile Components in Polymer Materials

In polymer materials, various additives and functional agents are added to improve the processing performance and service property. In addition, some low-molecular-weight components, such as various reaction and degradation products, are formed during polymerization, processing, and service. These components are present in quite low concentrations, but these concentrations significantly affect the property and stability of polymer materials. Most of the additives, functional agents, and low-molecular-weight components are volatile, which can be separated and analyzed by the flash evaporation of PGC.

Flash evaporation involves the evaporation of small volatile molecules from solid samples at relatively low temperatures (typically, 250–350°C). Simultaneously, the polymer backbone is not cleaved. By this technique, volatile components, such as solvents, plasticizers, antioxidants, light stabilizers, and flame retardants, are directly analyzed without pre-extraction or separation.

Nitrile rubber (NBR) is a good example. Typically, it is used as a seal rubber, with a complex composition. By flash evaporation, volatile components, including reactants of vulcanization package, antioxidants, plasticizers, and processing agents, are identified (Figure 2.31).

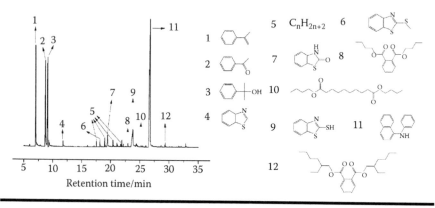

Figure 2.31 Result obtained from the flash evaporation of NBR and identification of volatile components.

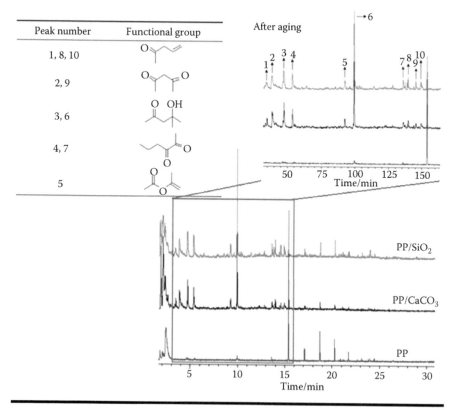

Figure 2.32 Volatile photo-oxidative degradation products obtained from aged PP nanocomposites. (From Yang, R. et al., *Polymer Degradation and Stability* 98, no. 12, 2013, 2466–2472.)

Another example is PP nanocomposites. When PP nanocomposites are exposed to the external environment, photo-oxidative degradation occurs under the action of ultraviolet light, oxygen, humidity, and heat, resulting in the scission of the chain and deterioration of mechanical properties. Before exposure, nearly no oxidative products are detected by flash evaporation. Nevertheless, in aged samples, various photo-oxidative products, such as ketones, alcohols, esters, and unsaturated species, are produced (Figure 2.32) [12].

2.4.8 Catalytic Degradation of Polymers

The recycling of waste plastics by pyrolysis to fuels or chemical raw materials is crucial for the environment as well as for sustainable development. A column filled with catalysts is connected between the pyrolyzer and GC for investigating the

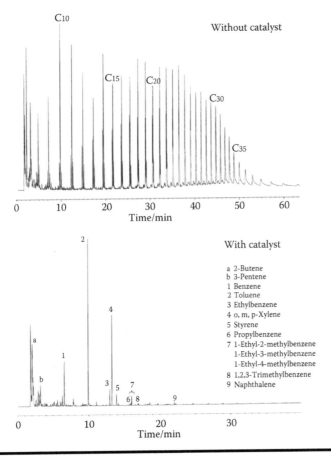

Figure 2.33 **Pyrograms of LDPE with and without catalysts. (From Serrano, D.P. et al., *Journal of Analytical and Applied Pyrolysis* 74, nos. 1–2, 2005, 370–378.)**

catalytic degradation of recycled polymers. For example, the pyrolysis of LDPE affords a series of saturated and unsaturated hydrocarbons. With solid acid catalysts, the pyrolysates are completely different. Large amounts of small hydrocarbons and aromatics, especially benzene, toluene, and xylene, are produced instead of long-chain hydrocarbons (Figure 2.33) [13].

References

1. Yang, R., X. Zhou, C. Luo, and K. Wang. *Advanced Instrumental Analysis of Polymers*, 3rd ed. Beijing: Tsinghua University Press, 2010.
2. Tsuge, S., H. Ohtani, and C. Watanabe. *Pyrolysis-GC/MS Data Book of Synthetic Polymers*. Amsterdam: Elsevier, 2011.
3. Tsuge, S., Y. Sugimura, and T. Nagaya. Structural characterization of polyolefines by pyrolysis-hydrogenation glass capillary gas chromatography. *Journal of Analytical and Applied Pyrolysis*, 1980, 1(3), 221–229.
4. Wampler, T.P. *Applied Pyrolysis Handbook*, 2nd ed. Boca Raton, FL: CRC Press, 2007.
5. Yang, R., S. Liu, and K. Wang. Standardization of PGC database and data sharing for polymer identification. *Chemistry*, 1999, 13, 99032.
6. Beynon, J.H.. *Mass Spectrometry and Its Application to Organic Chemistry*. Amsterdam: Elsevier Pub. Co., 1960.
7. Ohtani, H., T. Kimura, and S. Tsuge. Analysis of thermal degradation of terephthalate polyesters by high-resolution pyrolysis-gas chromatography, *Analytical Science*, 1986, 2(2), 179–182
8. Tsuge, S., D. Hiramitsu, T. Horibe, M. Yamaoka, and T. Takeuchi. Characterization of sequence distributions in methyl acrylate styrene copolymers to high conversion by pyrolysis-gas chromatography, *Macromolecules*, 1975, 8(6), 721–725.
9. Wang, F.C.-Y. The microstructure exploration of thermoplastic copolymer by pyrolysis-gas chromatography. *Journal of Analytical and Applied Pyrolysis*, 2004, 71, 83–106.
10. Wang, F.C.-Y., and P.B. Smith, Compositional and structural studies of vinylidene chloride/vinyl chloride copolymers by pyrolysis gas chromatography, *Analytical Chemistry*. 1996, 68(3), 425–430.
11. Ohtani, H., S. Tsuge, and T. Usami. Determination of short-chain branching up to C6 in low-density polyethylenes by high-resolution pyrolysis-hydrogenation gas chromatography. *Macromolecules*, 1984, 17, 2557–2561.
12. Yang, R., J. Zhao, and Y. Liu. Oxidative degradation products analysis of polymer materials by pyrolysis gas chromatography–mass spectrometry. *Polymer Degradation and Stability*, 2013, 98(12), 2466–2472.
13. Serrano, D.P., J. Aguado, J.M. Escola, J.M. Rodriguez, and G.S. Miguel. An investigation into the catalytic cracking of LDPE using Py-GC/MS. *Journal of Analytical and Applied Pyrolysis*, 2005, 74(1–2), 370–378.

Exercises

1. What is the principle of PGC? Why can it be utilized to study polymers?
2. Please compare PGC and GC and explain their similarities and differences.
3. In the pyrogram of PP, which structure is possible in the triads (C_9)?
4. What is the principle of organic mass spectrometry?
5. In a mass spectrum, will the peak with the highest m/e correspond to the molecular ion? Why?
6. What are the characteristics of the isotope group? How to identify isotopic peaks?
7. For interpreting a mass spectrum, which steps are to be followed?
8. If the molecular weight of a hydrocarbon is 142, and the intensity of the M+1 peak is 11% that of the M peak, please write the possible molecular formulas.

Chapter 3

Gel Permeation Chromatography

3.1 Introduction

Properties of polymers (particularly mechanical properties), processability, and solution characteristics are closely related to their molecular weights. Figure 3.1 plots the strength and processability of polymers as a function of their molecular weight. Before point A, the molecular weight was so low that the mechanical strength could not be determined, e.g., such as in the case of small organic molecules. After this point, strength significantly increased with molecular weight. Moreover, when the polymer molecular weight increased to point B, the polymer satisfied the requirements for its use in practical applications. From point B to C, the polymer strengthened slowly, increasing with molecular weight, while the processability significantly worsened. At this stage, it was quite difficult to process the polymer. In addition, the melt viscosity and elasticity of the polymer increased, while stress relaxation, crystallization speed, and solubility decreased. Hence, the molecular weight of the polymers must be controlled in a moderate range to attain satisfactory performance.

Since a polymer comprises several molecules with various molecular weights, the molecular weight is of an average value, and its distribution noticeably affects the polymer's properties. Consequently, the molecular weight of the polymer and its distribution are factors crucial for deciding the basic properties of polymer materials. Hence, it is imperative to determine the molecular weight of polymers and their distribution.

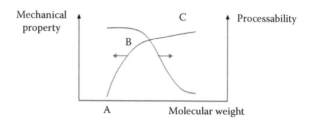

Figure 3.1 Mechanical property and processability as a function of molecular weight.

3.1.1 Definition of Average Molecular Weight

In contrast to small organic molecules with a definite molecular weight, polymers do not exhibit an exact molecular weight because they actually comprise molecules with different molecular weights. Hence, when discussing the molecular weight of polymers, the average value is typically utilized.

If there are N_i molecules in a polymer with a molecular mass of M_i, then their mass W_i is expressed as follows:

$$W_i = N_i M_i. \tag{3.1}$$

Several definitions of average molecular weights that are commonly utilized are explained as follows [1]:

Number-average molecular weight:

$$\bar{M}_n = \frac{\sum N_i M_i}{\sum N_i} = \frac{\sum W_i}{\sum W_i M_i^{-1}} \tag{3.2}$$

Weight-average molecular weight:

$$\bar{M}_w = \frac{\sum W_i M_i}{\sum W_i} \tag{3.3}$$

Z-average molecular weight:

$$\bar{M}_z = \frac{\sum Z_i M_i}{\sum Z_i} = \frac{\sum W_i M_i^2}{\sum W_i M_i} \tag{3.4}$$

Viscosity-average molecular weight:

$$\bar{M}_\eta = \left(\frac{\sum W_i M_i^\alpha}{\sum W_i} \right)^{1/\alpha} \tag{3.5}$$

Here, α is a parameter in the Mark–Houwink equation: $[\eta] = KM^\alpha$. When $\alpha = -1$, $\bar{M}_\eta = \bar{M}_n$; when $\alpha = 1$, $\bar{M}_\eta = \bar{M}_W$. Typically, α is 0.5–1; hence, $\bar{M}_n < \bar{M}_\eta \leq \bar{M}_W$, and $\bar{M}_\eta$ is similar to $\bar{M}_W$.

The measure of the distribution of the molecular weight in a given sample is expressed as the polydispersity index d, which is defined as the ratio of the weight-average molecular weight to the number-average molecular weight:

$$d = \bar{M}_W / \bar{M}_n \tag{3.6}$$

For a uniformly dispersed polymer, clearly, $d = 1$; on the other hand, for a polydisperse polymer, $d > 1$. Moreover, if the polydispersity index d is high, a wide molecular weight distribution is observed. Table 3.1 shows the range of the average molecular weights of common polymers.

The molecular weight distribution depends on the polymerization mechanism. For example, for polymers prepared by linear condensation, d is ca. 2; for polymers prepared by radical polymerization and terminated by radical–radical coupling, d is ca. 1.5; for polymers prepared by coordination polymerization, d is 8–30 [2]. Although d is a measure of the distribution width, it does not reflect the content of each fraction and the corresponding molecular weight. Figure 3.2 shows a differential or integral curve indicating the relationship between the fraction content and molecular weight.

3.1.2 Methods to Determine Molecular Weight and Distribution

Several methods can be employed to determine the average molecular weights of polymers. For example, by titration, the concentration of functional groups at the terminal of a polymer chain can be determined, which in turn can be utilized to calculate its average molecular weight. In addition, the average molecular weight can be determined according to the basic thermodynamic properties of a dilute polymer solution (e.g., by increasing the boiling point, decreasing the freezing point, and through osmotic pressure), through its dynamic properties (e.g., ultracentrifugation sedimentation velocity, viscosity, and volume exclusion), and its optical properties (light scattering). A majority of these methods are based on the colligative property of dilute polymer solutions. In a very dilute solution, independent

Table 3.1 Range of Average Molecular Weights of Common Polymers

Plastics	Average Molecular Weight/($\times 10^4$)	Fibers	Average Molecular Weight/($\times 10^4$)	Rubbers	Average Molecular Weight/($\times 10^4$)
High-density polyethylene	6–30	Polyethylene terephthalate	1.8–2.3	Natural rubber	20–40
Polyvinyl chloride	5–15	Polyamide 66	1.2–1.8	Styrene–butadiene rubber	15–20
Polystyrene	10–30	Polyvinyl alcohol	6–7.5	cis-1,4-Polybutadiene rubber	25–30
Polycarbonate	2–6	Cellulose	50–100	Chloroprene rubber	10–12

Source: Pan, Z. *Polymer Chemistry*, 4th ed. Beijing: Chemical Industry Press, 2007.

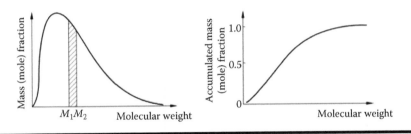

Figure 3.2 Molecular weight distribution curves: differential curve (left); integral curve (right).

polymer chains are present as random coils without any interaction. Hence, solution properties can be regarded as the sum of each molecule. These methods are based on different analytical principles and are calculated by different statistical methods; hence, their statistical implications are different, and these methods can be applied for polymers with a wide range of molecular weights (Table 3.2) [3].

These methods can provide information about the average molecular weights. To determine the molecular weight distribution, the various fractions must first

Table 3.2 Average Molecular Weight and Application of Various Methods

Method	Average Molecular Weight	Application Range
End-group analysis	$\overline{M}_n$	Less than 3×10^4
Increase of boiling point/ Decrease of freezing point	$\overline{M}_n$	Less than 3×10^4
Gaseous osmotic pressure	$\overline{M}_n$	Less than 2×10^4
Membrane osmotic pressure	$\overline{M}_n$	3×10^4 to 5×10^5
Viscosity	$\overline{M}_\eta$	2×10^4 to 10^6
Light scattering	$\overline{M}_W$	10^4 to 10^7
Ultracentrifugation sedimentation velocity	$\overline{M}_W$ and $\overline{M}_Z$	10^4 to 10^7
Small-angle x-ray scattering	$\overline{M}_W$	10^4 to 10^7
Electron microscopy	$\overline{M}_n$	Greater than 10^6
Gel permeation chromatography	$\overline{M}_n$, $\overline{M}_W$, $\overline{M}_Z$, and $\overline{M}_\eta$	Less than 10^7

Source: Zheng, C. *Molecular Weight and Distribution of Polymers*. Beijing: Chemical Industry Press, 1986.

be separated, and then the content and molecular weight of each fraction must be determined. Separation can be carried out by three methods:

(1) Separation based on the dynamic property of polymer chains in a solution: Ultracentrifugation sedimentation can be utilized for the determination of the average molecular weight and distribution. Various fractions are separated and determined according to different centrifugation sedimentation velocities.

(2) Separation based on the dependence of polymer solubility on molecular weight: The solubility of a high-molecular-weight polymer is less than that of a low-molecular-weight polymer. By increasing temperature, the solubility of a polymer increases. Hence, in a laboratory, the fractionation of polymers is typically conducted by precipitation or cooling. With the gradual addition of a certain amount of a precipitant into a polymer solution or with the decrease of temperature to a certain degree, the high-molecular-weight fraction is precipitated first. When the process attains equilibrium, this fraction can be separated. By continuously repeating this procedure, all fractions can be separated. By changing the amount of the precipitant in each fraction, we can obtain different numbers of fractions. The fractionation of polymers can also be realized by a reverse process, i.e., by the gradual addition of a solvent or an increase in temperature to dissolve low-molecular-weight fractions.

(3) The aforementioned methods are complex and time-consuming, and only discrete data can be obtained. A more appropriate method involves the determination of the molecular weight distribution based on the hydrodynamic volume of polymers via gel permeation chromatography (GPC), or size-exclusion chromatography (SEC) in some studies. GPC is discussed in this chapter.

3.2 Principle of GPC

GPC is a type of liquid chromatography. Different from other chromatographic methods, GPC involves the separation of molecules with various sizes, with no dependence on the interaction between the molecule and the mobile or stationary phase; nevertheless, it does rely on the different sizes of molecules in a dilute solution. Hence, GPC is a size-dependent separation method.

Similar to the other methods for determining molecular weights, GPC analysis is also based on the colligative property of the dilute polymer solutions. In GPC analysis, a polymer sample is dissolved in an appropriate solvent, affording a dilute solution. The solution concentration is sufficiently low to ensure that all molecule coils are independent and do not interact with each other. The larger the molecule weight, the larger the coil size.

A dilute polymer solution is passed through a GPC column by a pump. The GPC column, which is filled with porous filler particles of different pore sizes, permits the separation of molecules of different sizes. Pore size and distribution are the factors deciding the separation ability of the column. The total volume of the column V_t comprises of three volume components: a backbone volume of filler particles V_{CM}, the pore volume of the filler V_i, and an interfiller volume V_M (Equation 3.7). Among these parts, the backbone volume does not contribute to separation; hence, the separation is mainly realized by the hollow volume, i.e., V_i and V_M, where polymer molecules can enter.

$$V_t = V_{CM} + V_i + V_M.$$

(3.7)

Oversized molecules can only pass the column via the interfiller volume as their sizes are greater than all pore sizes. hence, their elution volume is equal to the interfiller volume:

$$V_R = V_M.$$

(3.8)

On the other hand, extremely small molecules can pass into all spaces in the column, as their sizes are less than all pore sizes. Their elution volume is equal to the sum of the interfiller volume and pore volume:

$$V_R = V_i + V_M.$$

(3.9)

For a given column, the interfiller volume, as well as the pore size and distribution, is definite; hence, the above two groups of molecules cannot be separated.

Moreover, molecules of intermediate size, i.e., not too large and not too small, can be successfully separated because molecules with different molecular weights have different coil sizes; hence, they can enter different portions of the pores:

$$V_R = KV_i + V_M.$$

(3.10)

Here, K, representing the ratio of the pore volume through which a molecule can enter to the total pore volume, is the partition coefficient. It represents the percentage of the volume occupied by a molecule, $0 \le K \le 1$. Various molecules with different molecular weights have different coil sizes, different K, and thus different exclusion volumes; hence, these molecules can be separated by GPC.

GPC exhibits the following features:

■ GPC can only determine a limited range of molecular weights. Molecules with extremely low and high molecular weights cannot be distinguished in a column; hence, their molecular weights cannot be determined.

- Molecules with high molecular weights are eluted first.
- The elution volume of molecules is not less than V_M nor greater than $V_i + V_M$.
- K depends on the molecular size, which is the hydrodynamic volume, or the volume of the equivalent sphere of a molecule.

3.3 Instrument and Sampling

3.3.1 Instrument

From the principle of GPC, a dilute polymer solution is passed through a column by a pump, during which molecules with different molecular weights are separated and detected. Hence, a GPC instrument (Figure 3.3) comprises four systems: flow system, separation system, detection system, and supplements.

1. Flow System

    The flow system consists of a solvent tank, vacuum degasser, and pump. The degasser removes bubbles from the mobile phase, while the pump ensures that the mobile phase or eluent constantly passes through the column, which affects GPC performance and data precision. Prerequisites for the pump include a small dead volume and steady flow rate. The flow rate can be easily adjusted in the range of 0.01–10 mL/min, with a precision of ±1%. Typically, a flow rate of 1 mL/min is utilized.

2. Separation System

    The separation system, which comprises an injector and a column, is the core part of a GPC instrument.

    Typically, the injector is a six-way valve (Figure 3.4). During injection, the sample ring is filled with a sample solution, but the sample ring is not connected to the mobile phase. When the injector is switched to the injection state, the valve turns such that the sample ring is connected to the mobile

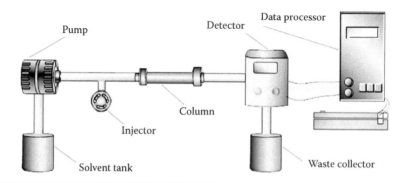

Figure 3.3   Schematic of a GPC instrument.

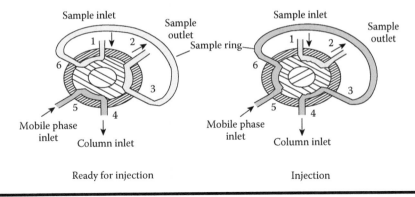

Figure 3.4 Schematic of a six-way valve for injection.

phase, and the sample solution in the ring is pushed by the eluent into the column. Typically, the volume of the sample ring is 50 or 100 μL. Such a design ensures precise and repetitive injection.

Typically, a stainless steel GPC column is 20–50 cm in length, with an inner diameter of 7–10 mm; it is filled with a porous gel exhibiting high porosity, high strength, a low swelling ratio, and a low thermal expansion coefficient. Typically, two types of gels are used [4]:

■ Crosslinked polystyrene is used to analyze oil-soluble polymers in organic solvents, with a series of monodisperse polystyrene as the standards.
■ Crosslinked PMMA or gelatin is used to analyze water-soluble polymers or biopolymers in water, with a series of narrowly dispersed polyethylene glycol (PEG) set as the standards.

As mentioned above, GPC involves the separation of molecules based on their hydrodynamic volumes: large molecules are eluted first, followed by small molecules; the elution volume does not exceed the total solvent volume that can be contained in a column; hence, the retention time range is predictable. It is feasible to inject at a certain time interval, when the overlapping of chromatographic peaks is not observed. An auto-injector has been designed accordingly, and the efficiency has been increased.

3. Detection Systems
　Typically, three detectors are utilized in GPC:
　(1)　Concentration Detector
　　　Two concentration detectors are available: refractive index (RI) detector and ultraviolet (UV) detector.
　　　Typically, a commercial GPC instrument is equipped with an RI detector, which operates on the basis of different refractive indices of the solvent and solution. The RI is a function of the polymer solution

concentration. Hence, the eluent concentration is continuously determined via the difference in the RIs of the sample solution and reference. The RI detector is a common detector, which can be utilized for detecting various polymer solutions. However, it does not exhibit high sensitivity. This detector is very sensitive to temperature changes; hence, it is crucial to carefully control the ambient temperature.

On the other hand, the UV detector exhibits high sensitivity, albeit only responding to some specific structures, such as conjugated or aromatic structures. Furthermore, as conjugated or aromatic structures are absent in the solvents typically used, the UV detector does not detect any signal from the mobile phase; thus, interference with the data is not observed.

(2) Viscosity Detector

The viscosity detector determines the intrinsic viscosity [η] of the eluted solution. The viscous-average molecular weight is calculated according to the Mark–Houwink equation:

$$[\eta] = KM^{\alpha}. \tag{3.11}$$

Here, K and α are constants, which are dependent on the polymer, solvent, and solution temperature. α is related to the molecular configuration. For a contracted structure, α is small. For a spherical molecule, α is nearly 0; for a random coil, α is 0.5–0.8; for a rigid rod, α may be greater than or equal to 1. If K and α are known, the molecular weight of the polymer can be calculated.

Typically, for intrinsic viscosity measurements, a four-capillary detector in a Wheatstone bridge arrangement (Figure 3.5) is utilized. R_1, R_2,

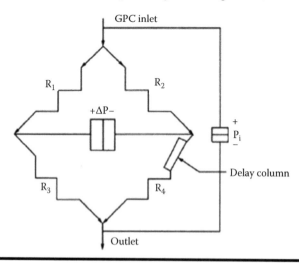

Figure 3.5   Schematic of the four-capillary viscosity detector.

R_3, and R_4 are the same capillaries. When solvent is passed through the detector, all four capillaries and the delay column are filled with the solvent. At this stage, the pressure difference ΔP is 0. When the sample solution enters through the detector, R_1, R_2, and R_3 are filled with the solution, while R_4 is still filled with the solvent, caused by the delay effect related to the large volume of the delay column. At this stage, ΔP is not 0; hence, the specific viscosity η_{sp} of the solution can be calculated from the pressure drop P_i between the inlet and outlet of the column and ΔP according to Equation 3.12.

$$\eta_{sp} = 4\Delta P/(P_i - \Delta P) \tag{3.12}$$

The intrinsic viscosity $[\eta]$ is calculated by the Solomon–Gatesman method:

$$[\eta] = \lim_{C \to 0} \frac{\eta_{sp}}{c} = \sqrt{2(\eta_{sp} - \ln(\eta_{sp} + 1)/c}. \tag{3.13}$$

Here, c is the solution concentration. As the eluted solution concentration is extremely low, the condition $c \to 0$ is considered true. In Equation 3.13, c is determined by the above-mentioned concentration detector. The viscosity detector requires a steady flow and a precisely controlled temperature because the pressure is sensitive to the flow rate, and viscosity is sensitive to the temperature.

(3) Molecular Weight Detector

A light scattering detector is utilized to determine the M_w of the eluted polymers; hence, it is also called a molecular weight detector.

When light is irradiated on a polymer molecule, Rayleigh scattering occurs, which can be expressed by the Rayleigh ratio R_θ:

$$R_\theta = r^2 I/I_0. \tag{3.14}$$

Here, I_0 and I represent the intensities of the incident light and scattering light, respectively; r is the distance between the scattering center and observer. The scattering light intensity depends on the scattering angle θ (the angle between the scattered and incident directions) and c. This relationship is also dependent on the weight, size, and conformation of polymer molecules, as described in Equation 3.15:

$$\left(\frac{1 + \cos^2 \theta}{2} \right) \frac{Kc}{R_\theta} = \frac{1}{MP_\theta} + 2A_2 c. \tag{3.15}$$

Here, K is the instrumental parameter, which depends on the incident light wavelength and RI of the solution, representing changes in the refractive index with c. M is the weight-average molecular weight. A_2 is the second virial coefficient, representing the interaction between chain segments and solvent molecules. P_θ is the scattering factor, which is related to the shape of the macromolecules and the incident light wavelength:

$$P_\theta = 1 - \frac{16\pi^2}{3(\lambda/n_0)^2} \overline{R^2} \sin^2 \frac{\theta}{2} + \dots \tag{3.16}$$

Here, λ is the incident light wavelength; n_0 is the refractive index of the solvent; and $\overline{R^2}$ is the mean-square gyration radius. For a random coil solution, $\overline{R^2} = \overline{h^2}/6$ ($\overline{h^2}$ is the mean-square end-to-end distance). Hence, the scattering equation of polymer solutions is described as follows:

$$\left(\frac{1 + \cos^2 \theta}{2} \right) \frac{Kc}{R_\theta} = \frac{1}{M} \left(1 + \frac{8\pi^2}{9} \frac{\overline{h^2}}{(\lambda/n_0)^2} \sin^2 \frac{\theta}{2} \right) + 2 A_2 c \tag{3.17}$$

If a series of scattering intensities R_θ at various concentrations and scattering angles are determined, and $\left(\dfrac{1 + \cos^2 \theta}{2} \right) \dfrac{Kc}{R_\theta}$ versus c is plotted, a series of $\left[\left(\dfrac{1 + \cos^2 \theta}{2} \right) \dfrac{Kc}{R_\theta} \right]_{c \to 0}$ at various scattering angles are obtained.

By plotting $\left[\left(\dfrac{1 + \cos^2 \theta}{2} \right) \dfrac{Kc}{R_\theta} \right]_{c \to 0}$ versus $\sin^2 \dfrac{\theta}{2}$, M can be obtained from the intercept, $\overline{h^2}$ and $\overline{R^2}$ can be calculated from the slope.

With the use of a low-angle laser light scattering detector (LALLS), $\theta \to 0$ can be considered true. Meanwhile, as the solution concentration in GPC is extremely low, it is also possible to consider that $c \to 0$. As a consequence, Equation 3.17 is simplified as follows:

$$\frac{Kc}{R_\theta} = \frac{1}{M} \tag{3.18}$$

Hence, by coupling a concentration detector with LALLS in GPC, c and R_θ can be directly determined, and M can be calculated.

By clearly coupling the viscosity detector with the molecular weight detector, $[\eta]$ and M can be directly determined, and the two important parameters K and α can be calculated.

Table 3.3   Common Solvents in GPC Analysis and Their Basic Properties

Solvent	Refractive Index (25°C)	Temperature Range/°C
Tetrahydrofuran (THF)	1.4040	25–50
Dimethylformamide (DMF)	1.4269	60–80
1,2,4-Trichlorobenzene	1.517	135–150

(4) Supplements

Supplements comprise a temperature control system, a data processing system, and a waste collector. As several detectors are sensitive to temperature changes, and characteristic parameters of polymers vary with temperature, a precisely controlled temperature is crucial.

3.3.2 Sampling

Dilute polymer solutions must be used as samples for GPC analysis. Hence, solvents exhibiting low viscosity, high boiling temperature, and good solubility, as well as low toxicity, are required. In addition, solvents cannot react with the gel in the column. With the use of the RI detector, it is imperative for the RI of the solvent to be different from that of the solution as far as possible for attaining high sensitivity (the RI of polymers is typically 1.4–1.6). On the other hand, with the use of the UV detector, it is imperative for the solvent to exhibit no absorbance or only weak absorbance. Table 3.3 summarizes the common solvents used in GPC analysis and their basic properties.

Typically, for GPC analysis, an extremely low solution concentration of 1–10 mg/mL is used. Polymer samples need to be purified before analysis for removing oligomers, emulsifiers, and monomers, so that the results are not affected by these small molecular components. Before injection, solutions are filtered to remove insoluble, dirty deposits. Polyolefins can only be dissolved at high temperatures; high-temperature GPC is utilized for their analysis.

3.4 Data Processing

3.4.1 GPC Chromatogram

Similar to other chromatograms, a GPC chromatogram plots the change in the signal intensity (y-axis) versus the retention value (elution volume or time in GPC). A typical GPC instrument comprises an RI detector; hence, the signal intensity reflects the elution concentration. The elution volume (x-axis) represents the molecular weight of a sample, which is dependent on the hydrodynamic volume

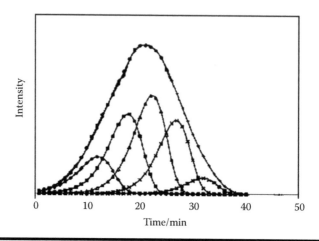

Figure 3.6 **GPC chromatogram of a polydisperse polymer sample.**

(molecular weight) of the molecular coil, while the y-axis represents the mass percentage of each fraction. Hence, a GPC chromatogram is actually the differential distribution curve of the molecular weight of a polymer.

For a monodisperse polymer sample, the chromatogram is expressed by a Gaussian function, and the peak value represents the molecular weight. On the other hand, for a polydisperse polymer sample, the chromatogram is considered to be the superposition of chromatograms of several monodisperse fractions, as shown in Figure 3.6. The shape of the curve depends on the molecular weight distribution; hence, it is not necessarily a Gaussian function, and the peak value possibly does not reflect the average molecular weight. In this case, the average molecular weight and its distribution cannot be directly determined. Hence, such data are further subjected to data processing procedures.

3.4.2 Molecular Weight Calibration Curve

A GPC chromatogram plots the elution concentration versus the elution volume. The relationship between the elution volume and molecular weight can be utilized to calculate the average molecular weight and distribution. The log M versus V curve is referred to as the molecular weight calibration curve, the precision of which is a decisive factor for that of the determined molecular weight.

A calibration curve can be obtained by three methods: using narrow-dispersed standards, using polydisperse standards, and the universal calibration method.

1. Calibration Curve Using Narrow-dispersed Standards
   For an unknown sample, if a series of known narrow-dispersed ($d < 1.1$) standards of the same polymer are obtained, and GPC analysis of these

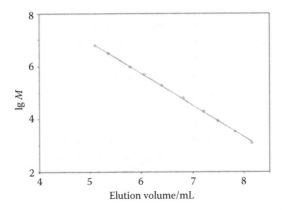

Figure 3.7   Narrow-dispersed standard calibration curve.

standards is carried out to determine the elution volume, the log *M–V* curve
can be plotted as shown in Figure 3.7:

$$\lg M = A - BV. \tag{3.19}$$

Here, *A* and *B* are constants, which can be obtained by curve fitting.

This method is quite easy and exhibits high precision. However, it is often
difficult to obtain the narrow-dispersed polymer standards with known
molecular weights, which is a considerable limitation.

From the separation principle discussed in Section 3.2, for a column, a
limited range of molecular weights can be determined. Hence, the calibration
curve is not a line, but a segment, as shown in Figure 3.8. Point A is referred
to the exclusion limit, which corresponds to the highest molecular weight
that can be determined. Molecules with a molecular weight greater than that
corresponding to point A cannot be distinguished. On the other hand, point
B is referred to the permeation limit, which corresponds to the lowest molecular weight that can be determined. Molecules with a molecular weight less
than that corresponding to point B cannot be distinguished. Only molecules
with molecular weights greater than those corresponding to point B and less
than those corresponding to point A can be separated and distinguished. As
the molecular weights approach either point A or B, the rate of measurement
error increases.

2. Calibration Curve Using Polydisperse Standards

In most cases, it is difficult to obtain narrow-dispersed polymer standards for determination. Even so, a calibration curve can still be obtained by
iteration using 2–3 polydisperse standards (with known average molecular
weights). Hence, it is not crucial to use narrow-dispersed standards.

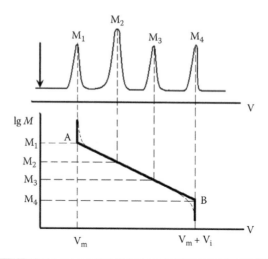

Figure 3.8 Separation limits of GPC.

First, the GPC analysis of polydisperse standards must be carried out, and GPC chromatograms must be recorded. Second, the values of A and B in Equation 3.19 must be randomly set for constructing a calibration curve. Third, the average molecular weights of these standards must be calculated according to the calibration curve, and the results must be compared with known data. If the difference is greater than a preset error (typically 5–10%), the second and third steps must be repeated. This procedure must be repeated until the difference between the calculated result and the known data is less than the error. Then, A and B values are set, and a calibration curve is obtained.

This method is advantageous as it is easy and does not require the use of narrow-dispersed polymer standards. However, the exclusion and permeation limits cannot be determined by this method, and its precision is not very high.

3. Universal Calibration Method

GPC provides information linking the elution volume to the hydrody-namic volume of polymer molecules. Various polymer molecules exhibit dif-ferent flexibilities. For polymer molecules with the same molecular weight but different configurations, their hydrodynamic volumes in solution are different. Hence, the calibration curves determined by the above two meth-ods can only determine the same polymer as the standard. When the poly-mer to be determined is changed, a new calibration curve is required, which is not convenient. If the calibration curve is determined on the basis of the hydrodynamic volume, it is universal and can be utilized for various polymers.

According to the equivalent hydrodynamic sphere model, the Einstein viscosity equation is expressed as follows:

$$[\eta] = 2.5NV/M. \tag{3.20}$$

Here, $[\eta]$ is the intrinsic viscosity; M is the molecular weight; V is the hydrodynamic volume of the equivalent sphere of a polymer molecule; and N is the Avogadro constant. Hence, $[\eta]M$ can be used to express the hydrodynamic volume of a polymer molecule.

The $\lg[\eta]M$–V curve is more universal when compared to the $\lg M$–V curve. That is, if two polymers exhibit the same elution volume under the same GPC condition, then:

$$[\eta]_1 M_1 = [\eta]_2 M_2. \tag{3.21}$$

Subscripts 1 and 2 represent two polymers. By substituting the Mark–Houwink equation (Equation 3.11 into Equation 3.21):

$$K_1 M_1^{1+\alpha_1} = K_2 M_2^{1+\alpha_2}. \tag{3.22}$$

By taking the logarithm on both sides:

$$\lg K_1 + (1+\alpha_1)\lg M_1 = \lg K_2 + (1+\alpha_2)\lg M_2$$
$$\lg M_2 = \frac{1}{1+\alpha_2}\lg\left(\frac{K_1}{K_2}\right) + \frac{1+\alpha_1}{1+\alpha_2}\lg M_1 \tag{3.23}$$

Hence, when K_1, α_1, K_2, and α_2 under the given experimental conditions are known, the calibration curve of the second polymer can be calculated from that of the first polymer, according to Equation 3.23.

Previous studies have reported that this method is applicable for linear and random coil polymers [5]. The applicability of this method for long-branched polymers, or rigid polymers, is under investigation.

The calibration curves of several polymers can be obtained from the calibration curve of one polymer (typically narrow-dispersed polystyrene). The prerequisite is that K and α of these two polymers should be known. Table 3.4 summarizes the K and α values of common polymers in conventional solvents.

Table 3.4 Parameters of Common Polymers in Solvents

Solvent	Polymer	Temperature/°C	$K \times 10^3/(mL/g)$	α	Molecular Weight $\times 10^{-4}$
Tetrahydrofuran (THF)	PS	25	1.60	0.706	> 0.3
		23	68.0	0.766	5–100
	PS (comb)	23	2.2	0.56	15–1120
	PS (star)	23	0.35	0.74	15–60
	PVC	23	1.63	0.766	2–17
	PMMA	23	0.93	0.72	17–130
	Polycarbonate (PC)	25	3.99	0.77	
	Polyvinyl acetate (PVAc)	25	3.5	0.63	1–100
	Polyisoprene (PIP)	25	1.77	0.735	4–50
	Natural rubber (NR)	25	1.09	0.79	1–100
	Polybutadiene rubber (BR)	25	0.85	0.75	0.4–400
	1,4-PB	25	76	0.44	27–55

(Continued)

Table 3.4 (Continued) Parameters of Common Polymers in Solvents

Solvent	Polymer	Temperature/°C	$K \times 10^3/(mL/g)$	α	Molecular Weight $\times 10^{-4}$
o-Dichlorobenzene	PS	135	1.38	0.7	0.2–90
m-Cresol	PE	135	4.77	0.7	0.6–70
Chloroform		135	5.046	0.693	1–100
	PP	135	1.3	0.78	2.8–46
	PS	135	2.02	0.65	0.4–200
	Polyethylene terephthalate (PET) fiber	135	1.75	0.81	0.27–3.2
	PS	25	7.16	0.76	12–280
		25	11.2	0.73	7–150
		30	4.9	0.794	19–373
	PVAc	25	20.3	0.72	4–34
	Polyvinylpyrrolidone (PVP)	25	19.4	0.64	2–23
	PMMA	25	4.8	0.8	8–137
		30	4.3	0.8	13–263

(Continued)

Table 3.4 (Continued)   Parameters of Common Polymers in Solvents

Solvent	Polymer	Temperature/°C	$K \times 10^3/(mL/g)$	α	Molecular Weight $\times 10^{-4}$
	Polybutyl methacrylate (PBMA)	25	4.37	0.8	8–80
	PC	25	11	0.82	0.8–27
	Polyethyl acrylate (PEA)	30	31.4	0.68	9–54
	Polyethylene oxide (PEO)	25	206	0.5	< 0.15
	Ethyl cellulose	25	11.8	0.89	4–14
Acetone	PMMA	25	7.5	0.70	2–740
	PBMA	25	18.4	0.62	100–600
	Polymethyl acrylate (PMA)	25	5.5	0.77	28–160
	PEA	30	20	0.66	16–50
	Polyisopropyl acrylate (PiPA)	30	13	0.69	6–30
	PBA	25	6.85	0.75	5–27
	PEO	25	156	0.5	< 0.3
	Polymethyl acrylonitrile (PMAN)	20	95.5	0.53	35–100
	NBR	25	50	0.64	2.5–100

3.4.3 Molecular Weight and Distribution

For monodisperse samples, the elution volume can be obtained from the GPC chromatogram, and the molecular weight can be directly calculated from the calibration curve.

On the other hand, for polydisperse samples, two methods can be employed to calculate the molecular weight and distribution of the polymer—function and integral methods—respectively.

1. Function Method

   The function method is based on the principle that if a function for expressing a GPC curve can be found, various average molecular weights of polymers can be calculated accordingly. In most cases, a GPC curve is symmetric, which can be expressed by a Gaussian function. The mass percentage of a fraction can be described as follows:

$$W(V) = \frac{1}{\sigma\sqrt{2\pi}} \exp\left[-\frac{(V-V_p)^2}{2\sigma^2}\right]. \tag{3.24}$$

   Here, V and V_p represent the elution volume of the fraction and the peak elution volume, respectively. For obtaining the average molecular weight, the elution volume V should be converted to the molecular weight M. For convenience, Equation 3.19 is converted as follows:

$$\ln M = A - B_1 V_e. \tag{3.25}$$

   Here, slope $B_1 = 2.303B$. At the same time, Equation 3.26 should be satisfied (this implies that the total mass is constant):

$$\int_0^\infty W(V)\,dV = \int_0^\infty W(M)\,dM. \tag{3.26}$$

   Hence, the mass differential distribution function with molecular weight as the variable is described as follows:

$$W(M) = \frac{1}{M\sigma'\sqrt{2\pi}} \exp\left[-\frac{1}{2}\left(\frac{\ln M - \ln M_p}{\sigma'}\right)^2\right]. \tag{3.27}$$

Here, $\sigma' = B_1\sigma$; where M_P is the peak molecular weight, which can be calculated from the calibration curve. Then, $\bar{M}_w$, $\bar{M}_n$, and d can be calculated from Equations 3.28 to 3.30, respectively:

$$\bar{M}_w = M_P \exp\left(B_1^2\sigma^2/2\right) \tag{3.28}$$

$$\bar{M}_n = M_P \exp\left(-B_1^2\sigma^2/2\right) \tag{3.29}$$

$$d = \bar{M}_w/\bar{M}_n = \exp\left(B_1^2\sigma^2\right). \tag{3.30}$$

Hence, if a GPC curve is simulated by a Gaussian function, the average molecular weights and d only depend on M_P, slope B_1 from the calibration curve, and the width σ of the GPC peak. If a GPC curve is asymmetric or consists of multiple peaks, a Gaussian function is not suitable for simulation. In this case, the integration method can be employed to calculate the average molecular weight.

2. Integration Method

A GPC curve can be divided to n slices, representing n fractions in a sample. Each fraction is considered to be monodisperse. The elution time t_i and concentration H_i of the fraction i can be obtained from the GPC curve. The molecular weight M_i can be calculated from the calibration curve. Hence, the mass percentage W_i of the fraction i is expressed as follows:

$$\bar{W}_i = H_i / \sum_{i=1}^{n} H_i. \tag{3.31}$$

Then, the number- and weight-average molecular weights are calculated from Equations 3.32 and 3.33, respectively, as follows:

$$\bar{M}_n = \frac{1}{\sum \bar{W}_i/M_i} = \frac{\sum H_i}{\sum H_i/M_i} \tag{3.32}$$

$$\bar{M}_W = \sum \bar{W}_i M_i = \frac{\sum H_i M_i}{\sum H_i} \tag{3.33}$$

In addition to $\bar{M}_n$, and $\bar{M}_w$, $\bar{M}_z$, $\bar{M}_\eta$, and d values of polymers can also be calculated. The integration method is advantageous as GPC curves with various shapes can be calculated; furthermore, it is quite easy to be conducted using a computer; hence, it is the main data processing method in GPC.

3.5 Applications

The molecular weight and distribution of polymers are basic parameters for determining various properties. GPC is powerful and convenient for the determination of the molecular weight and distribution of polymers; hence, it is one of the most important analytical methods for the characterization of polymer materials. GPC can be employed to investigate all processes involving changes in molecular weight and distribution, such as polymerization, processing, and aging. In addition, polymer configuration (such as branching), and conformational changes in solutions (such as coil size), can be investigated with the use of a light scattering or viscosity detector in the GPC system.

3.5.1 Structural Analysis

The molecular weight and distribution of polymers determine their properties. Polymers with various average molecular weights and distributions can be prepared using various catalysts and polymerization methods. For example, three types of polyethylene—high-density polyethylene (HDPE), low-density polyethylene (LDPE), and linear low-density polyethylene (LLDPE)—are with different branching degrees. HDPE is a linear macromolecule, with high crystallinity, high strength, and relatively low ductility. LDPE is composed of several branches, with relatively low crystallinity, low strength, but good ductility. LLDPE is composed of moderate branches. Hence, LDPE and LLDPE are often used for preparing films, e.g., greenhouse films, while HDPE can be used to prepare pipes. In addition, HDPE with a bimodal distribution is typically utilized to prepare high-performance pipes. The high-molecular-weight portion ensures high strength, high melt strength, and low melt deformation, while the low-molecular-weight portion ensures the improved processability of pipes. Figure 3.9 shows the typical elution curves of polyethylenes (PEs) with various distributions.

In addition to the average molecular weight and distribution, branching is another factor affecting polymer properties. Two types of branches—long branches (with greater than or equal to six carbon atoms) and short branches (with one to five carbon atoms)—are present. The presence of short branches destroys the regularity of macromolecular chains and interferes with crystallization, while the presence of long branches significantly increases the zero-shear viscosity and affects processing, albeit their concentration is extremely low.

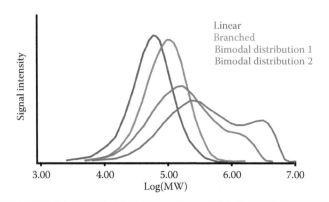

Figure 3.9 Typical elution curves of polyethylenes with various molecular weight distributions.

The branching factor G is defined as the ratio of intrinsic viscosities of the branched and linear polymers ($[\eta]_B$ and $[\eta]_L$, respectively) with the same molecular weight:

$$G = g^\varepsilon = [\eta]_B / [\eta]_L . \qquad (3.34)$$

Here, g is the branching index, which represents the ratio of the mean-square gyration radii of the branched and linear polymers with the same molecular weight. ε, which is typically 0.5–1.5, is a parameter related to the type of branching.

As compared to linear polymers with the same molecular weight, branched polymers exhibit lower intrinsic viscosity and a smaller hydrodynamic volume in the same solvent. Hence, in GPC analysis, as compared to linear polymers, branched polymers exhibit a longer elution time.

If a branched polymer has the same hydrodynamic volume as a linear polymer, then:

$$[\eta]_B M_B = [\eta]_L M_L . \qquad (3.35)$$

The molecular weight of a linear polymer M_L can be determined by GPC. $[\eta]_L$ can be calculated according to the Mark–Houwink equation (Equation 3.11; when K and α are known). If a GPC instrument is equipped with a light scattering detector (typically as an option), the molecular weight of a branched polymer M_B can be determined, and then $[\eta]_B$ can be calculated according to Equation 3.35. Then, G is obtained. If a GPC instrument is equipped with a viscosity detector (typically as an option), $[\eta]_L$ and $[\eta]_B$ can be directly determined, and G can be directly calculated according to Equation 3.34.

3.5.2 *Polymerization and Processing*

During the polymerization of monomers, the molecular weight continuously increases. For different polymerization methods, the molecular weight increases via different modes. For example, in radical polymerization, a high-molecular-weight fraction appears at the initial stage, albeit with extremely low conversion. The concentration of the high-molecular-weight fraction, as well as conversion, increases with time. In step polymerization, the conversion is quite high at the initial stage, but the average molecular weight is extremely low. The average molecular weight gradually increases with time. Both modes can be clearly distinguished by GPC analysis. Figure 3.10 shows the evolution of elution curves during radical polymerization and step polymerization.

During processing, the molecular weight may change because of high temperature and shear; hence, the mechanical properties may also change. Table 3.5 lists the average molecular weights of four polycarbonates (PCs) before and after processing. Different extents of degradation are observed. Sample PC-D exhibits the highest weight-average molecular weight; hence, it is expected to exhibit the highest impact resistance. However, this is not the case. By its molecular weight distribution, PC-D contains a high percentage of the low-molecular-weight fraction (less than 2×10^4). With the decrease of the molecular weight of PC to less than 2×10^4, all of its properties significantly deteriorate. Hence, impact strength decreases with the increasing content of the low-molecular-weight fraction, although processability is improved.

The optimal processing conditions of a polymer can be determined by sampling and analysis during processing. For example, various types of raw rubbers exhibit different changes in their molecular weight distribution during plasticization; hence, the optimal plasticization time can be decided from GPC analysis. Natural rubber (NR) comprises nonsoluble gels, which are not detected by GPC. With increasing plasticization time, GPC curves indicate the decrease of the average molecular weight; however, small peaks corresponding to high-molecular-weight polymers are observed, indicating gel breakage. With a longer plasticization time, these small peaks gradually disappear, indicating the further decrease of the average molecular weight, as well as a narrower distribution. After this point, the distribution remains nearly constant.

3.5.3 *Aging of Polymer Materials*

Polymer materials are often affected by various atmospheric factors (such as UV light, heat, oxygen, humidity, and microbes) and various service conditions (such as high or low temperature; pressure; medium; and irradiation, during storage and applications). As a result, macromolecular chains may be subjected to breakage, oxidation, or crosslinking; hence, their performances gradually deteriorate until the polymer materials fail at last. This process is known as aging. As all changes occurring during

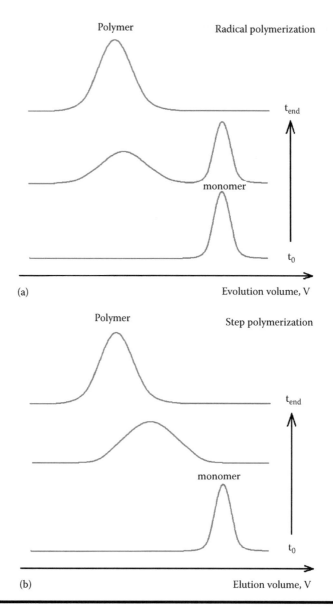

Figure 3.10 Schematic of the GPC elution curves of products during radical polymerization and step polymerization.

Table 3.5 Molecular Weights of Four Polycarbonates Before and After Processing

Sample	PC-C		PC-T		PC-S		PC-D	
	Before	After	Before	After	Before	After	Before	After
MW×10⁴								
$\overline{M}_W$	3.30	3.22	3.64	3.06	2.58	2.50	3.58	3.24
$\overline{M}_n$	1.40	1.40	1.45	1.21	1.18	1.14	1.15	1.03
$\overline{M}_z$	4.87	4.79	5.62	4.78	3.91	3.83	7.27	6.52
$\overline{M}_\eta$	3.16	3.08	3.48	3.06	2.46	2.39	3.32	3.02
Distribution								
Greater than 4×10^4	31.3%	29.9%	36.2%	27.5%	19.3%	18.1%	30.5%	28.8%
2×10^4 to 4×10^4	36.3%	36.2%	32.9%	34.2%	35.2%	34.7%	26.7%	28.5%
Less than 2×10^4	32.4%	33.9%	30.9%	38.3%	45.5%	47.2%	42.8%	44.7%

aging are accompanied by the change of the molecular weight and distribution, GPC is a powerful tool for investigating the aging behavior and mechanism.

Polyesters can be easily hydrolyzed as ester bonds can undergo simple hydrolysis to their corresponding acids and alcohols. For example, PC is a good engineering plastic, but its resistance to water, particularly hot water, is very weak. Hence, blending PC with PE can significantly improve its water resistance. Figure 3.11 shows the changes in the molecular weight of PC and the PC/PE blend with aging time in hot water (100°C and 80°C) by GPC.

In 100°C hot water, the molecular weight of PC most rapidly decreased. After approximately only 20 days, the average molecular weight decreased to below 20,000. As mentioned before, PC with such a low molecular weight cannot be used as an engineering plastic. When PC was blended with PE, the original molecular weight marginally decreased because of processing-induced degradation. As compared to PC, the PC/PE blend exhibited a significantly slower rate of hydrolysis.

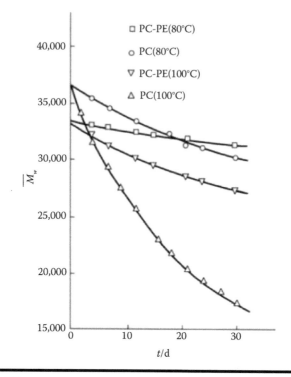

Figure 3.11 $\overline{M_w}$-*t* curves of PC and the PC/PE blend aged in hot water of 100° and 80°.

During hydrolysis, the molecular weight distributions of PC and the PC/PE blend are approximately the same.

From the perspective of hydrolysis dynamics, in most cases, the hydrolysis of ester bonds is a random process, with the random scission of chains. The bond breakage number n_t is expressed as follows:

$$n_t = \frac{1}{\overline{M_{nt}}} - \frac{1}{\overline{M_{n0}}}. \tag{3.36}$$

Here, $\overline{M_{n0}}$ and $\overline{M_{nt}}$ are the number-average molecular weights at the beginning and at reaction time *t*, respectively. At *t*, the probability α of a covalent bond to break is expressed as follows:

$$\alpha = n_t / (DP_0 - 1). \tag{3.37}$$

Here, DP_0 is the original degree of polymerization. When $DP_0 \gg 1$:

$$\alpha = n_t / DP_0. \tag{3.38}$$

By substituting Equation 3.36 in Equation 3.38:

$$\alpha = 254\left(\frac{1}{M_{nt}} - \frac{1}{M_{n0}}\right). \tag{3.39}$$

Here, 254 corresponds to the molecular weight of the repeating unit of PC. The hydrolysis rate constant K of PC is expressed as follows:

$$K = d\alpha/dt. \tag{3.40}$$

When the degree of hydrolysis is low, and only a few bonds are broken, hydrolysis can be regarded as a first-order reaction. Hence,

$$Kt = 254\left(\frac{1}{M_{nt}} - \frac{1}{M_{n0}}\right). \tag{3.41}$$

During the hydrolysis of PC, d is approximately unchanged, and the $\overline{M_W}$ determined is more precise; hence, the above equation can also be expressed as follows:

$$\frac{1}{M_{wt}} - \frac{1}{M_{w0}} = \frac{K}{254d}t. \tag{3.42}$$

Plotting $\left(\dfrac{1}{M_{wt}} - \dfrac{1}{M_{w0}}\right)$ versus t, we can obtain Figure 3.12. The degradation rate constant K can be calculated from the slope, as shown in Table 3.6. A good linear relationship is demonstrated by the first-order reaction mechanism for the hydrolysis of PC. By the Arrhenius equation, the activation energy for the hydrolysis of PC is calculated to be $E = (80 \pm 8)$ kJ/mol.

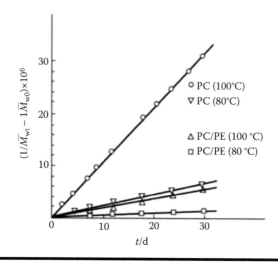

Figure 3.12   Degradation rates of PC and the PC/PE blend in hot water.

Table 3.6   Hydrolysis Rate Constants of PC and the PC/PE Blend at Various Temperatures

Sample	PC		PC/PE	
Temperature/°C	100	80	100	80
K/(mol/(mol·s))	68.30	13.63	14.14	3.38

References

1. He, M., H. Zhang, W. Chen, and X. Dong. *Polymer Physics*, 3rd ed. Shanghai: Fudan University Press, 2010.
2. Pan, Z. *Polymer Chemistry*, 4th ed. Beijing: Chemical Industry Press, 2007.
3. Zheng, C. *Molecular Weight and Distribution of Polymers*. Beijing: Chemical Industry Press, 1986.
4. Zhang, L., Q. Xue, Z. Mo, and X. Jin. *Modern Analytical Methods of Polymer Physics*, 2nd ed. Wuhan University Press, 2006.
5. Tung, L. H. *Fractionation of Synthetic Polymer: Principles and Practices*, New York: M. Dekker, 1977.

Exercises

1. What is the mechanism of separation by GPC?
2. Why is calibration required when GPC is employed for the determination of the average molecular weights and distributions of polymers? What are

the different calibration methods? Please compare their advantages and disadvantages.

3. If the elution time of the solvent is 15 minutes by GPC, can we inject another sample 15 minutes later? Why?

4. What are the requirements for a sample to be used for GPC analysis?

5. When the weight-average molecular weight $\overline{M}_w$ of a sample is determined using a small-angle light scattering detector, what assumptions are made? Can these assumptions be satisfied in actual cases?

6. K and α values of PS and PMMA at 30°C in a THF solution have been provided:
   PS: $K = 0.000128$ and $\alpha = 0.712$
   PMMA: $K = 0.000128$ and $\alpha = 0.690$
   Retention times and molecular weights of PS have been provided:

Retention time (min)	18.222	19.72	20.206	21.41	23.523	25.357	26.553	27.594
MW of PS	240000	110000	100000	50000	17500	8500	4000	1800

   Please calculate the calibration curve equations for PS and PMMA.

7. Molecular weight changes of PC hydrolyzed in 100°C hot water are shown in the table below:

Time (day)	0	1	2	4	5	9	7	12
$\overline{M}_n$	17516	13066	14071	11130	11131	9896	10577	9123
$\overline{M}_W$	39095	32756	31340	28883	29197	25429	27259	24613
Time (day)	16	18	21	24	27	30	35	
$\overline{M}_n$	6896	6988	6616	5778	5859	7093	5719	
$\overline{M}_W$	19930	20142	18346	17440	16495	18878	14615	

   Please calculate the hydrolysis dynamic equation of PC.

8. A linear polymer and a branched polymer with the same molecular weight are analyzed by GPC using the RI, viscosity, and light scattering detectors. Is there any difference among the signals observed by these detectors? Why? If yes, what is the difference?

Chapter 4

Spectroscopy

4.1 Introduction

As the name suggests, spectral analysis or spectroscopy involves the study of the interaction between matter and light at different frequencies (light spectrum); this interaction results in a change of the movement of atoms or molecules in matter, leading to energy transitions.

Spectral analysis methods are broadly categorized into three types: absorption spectroscopy (e.g., infrared and ultraviolet spectroscopy), emission spectroscopy (e.g., fluorescence spectroscopy), and diffraction spectroscopy (e.g., Raman spectroscopy). Let us assume a sample is in the ground state. When the sample is irradiated with light, it absorbs light energy and moves from a low-energy level to a high-energy level, resulting in an absorption spectrum. Conversely, when the sample returns from a high-energy level to a low-energy level, it emits light energy, resulting in an emission spectrum. In another case, the sample does not absorb or emit light, but exhibits inelastic collision with light, resulting in light diffraction. The sample also absorbs light energy and moves to a high-energy level. Part of the light energy is lost and its frequency changes, resulting in a diffraction spectrum.

In this chapter, only the following basic analytical methods are described: ultraviolet–visible (UV–Vis), infrared (IR), and Raman spectroscopy.

Matter is composed of molecules. A molecule exhibits three types of movement: the movement of electrons around the nucleus with electronic energy (E_e); the vibration and rotation of atoms with vibrational energy (E_v) and rotational energy (E_r), respectively. The energy difference between the electronic levels, ΔE_e, is the highest (1–20 eV). The energy difference between the rotation levels, ΔE_r, is the lowest (10^{-4} to 0.05 eV). The energy difference between vibration levels, ΔE_v, is moderate (0.05–1 eV). Thus, when moving from a low-energy level to a high-energy level, different amounts of energy are required for different types of movement, corresponding to the light of different frequencies. For example, when a sample is irradiated with UV–Vis light (λ = 200–800 nm), electronic transitions occur because the energy is in the ΔE_e range. On the other hand, when a sample is irradiated by mid-IR light (λ = 2.5–25 μm), vibrational and rotational transitions occur because the energy is in the ΔE_v and ΔE_r ranges, respectively. Moreover, when the sample is irradiated with microwave or far-IR light (λ = 0.1–1 mm), only rotational transitions occur. When a sample is irradiated with various frequencies, diverse information can be obtained about its molecular structure; hence, qualitative and quantitative analysis is possible based on the obtained information.

When a sample is irradiated with light of a wide range of wavelengths, some of the light is absorbed by the sample, which decreases the transmitted light energy. Hence, the absorption spectrum is a measure of the intensity of the transmitted light with the frequency (wavelength). The horizontal axis represents the frequency or wavelength of light, corresponding to the jump in the vibrational energy level; the absorption spectrum provides information about molecular structures and is the basis of qualitative analysis. The vertical axis represents the light intensity; absorption obeys the Beer's Law, i.e., absorption intensity is proportional to the number of photons absorbed by a sample, which is dependent on the possibility of the energy jump and the number of molecules. Hence, light intensity is the basis of quantitative analysis. Typically, light intensity is expressed by transmittance T (%) or absorbance A:

$$T\,(\%) = 100 \times I/I_0 \tag{4.1}$$

Here, I_0 and I represent the intensities of the incident and transmitted light, respectively.

$$A = \lg(I_0/I) = \varepsilon c l \tag{4.2}$$

Here, ε is the molar absorptivity (L/mol·cm), c is the molar concentration (mol/L), l is the sample thickness or optical path length (cm). Equation 4.2 is also referred to as the Lambert–Beer's law, which is the basis for the quantitative analysis of the absorption spectrum. The Lambert–Beer's law is only valid for dilute solutions. At high concentrations, ε is probably not constant and changes with concentration, leading to deviations.

A spectrum can be represented in two ways: with either transmittance or absorbance as the vertical axis. For example, Figure 4.1 shows the IR spectra of polystyrene (PS).

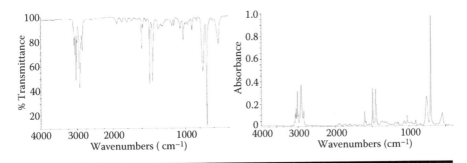

Figure 4.1   Infrared spectra of polystyrene with transmittance (left) and absorbance (right) as the vertical axis.

Absorbance follows the addition law. When more than one component contributes to the same absorbance and there is no interaction among these components, the total absorbance equals the sum of the absorbance of each component:

$$A_{total}^{\lambda} = A_1^{\lambda} + A_2^{\lambda} + A_3^{\lambda} + ... + A_n^{\lambda}$$
$$= (\varepsilon_1 c_1 + \varepsilon_2 c_2 + \varepsilon_3 c_3 + ... + \varepsilon_n c_n)l \qquad (4.3)$$

A spectrum of a polymer provides two types of information. One is obtained from the repeating units, where peaks, referred to as elementary peaks, are similar to those of corresponding organics. The other is obtained from the interactions among adjacent functional groups, including macromolecule–solvent interaction, intermacromolecular interaction, and intramacromolecular interaction, where peaks correspond to the alignment of segments and the aggregate structure of macromolecules. These peaks are referred to as polymeric peaks, which reflect polymer characteristics [1].

4.2 Ultraviolet–Visible Spectroscopy

UV–Vis spectroscopy involves the study of the interactions between samples (typically organics) and ultraviolet (wavelength range of 200–400 nm) and visible light (wavelength range of 400–800 nm). Light energy corresponds to electronic transitions [2].

When a molecule absorbs a photon, an electron from the outer layer will jump from the ground state to the excited state. Molecules with various chemical structures can absorb light of different wavelengths and exhibit different electronic transitions, with various possibilities. Hence, it is possible to identify various structures according to the absorption wavelength and intensity.

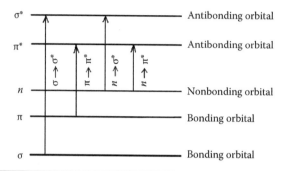

Figure 4.2 Energy levels and electronic transitions.

4.2.1 Electronic Transitions

When organics absorb UV–Vis light, four electronic transitions are possible: $\sigma \rightarrow \sigma^*$, $n \rightarrow \sigma^*$, $\pi \rightarrow \pi^*$, and $n \rightarrow \pi^*$ (Figure 4.2). Generally, electronic transition energy follows the order of $\sigma \rightarrow \sigma^* > n \rightarrow \sigma^* > \pi \rightarrow \pi^*$ (conjugated) $> n \rightarrow \pi^*$.

1. $\sigma \rightarrow \sigma^*$ transition

    Typically, saturated hydrocarbons are composed of C–C σ bonds. A significant amount of energy is required for the $\sigma \rightarrow \sigma^*$ transition, which corresponds to the absorption of light with a wavelength of less than 150 nm. Hence, as saturated hydrocarbons do not exhibit peaks in a typical UV–Vis spectrum, they can be used as solvents for samples.

2. $n \rightarrow \sigma^*$ transition

    Atoms such as oxygen, nitrogen, and sulfur, as well as halogens, have unshared pairs of electrons. For a saturated hydrocarbon molecule containing these atoms, $n \rightarrow \sigma^*$ transitions are possible. This transition requires significant amounts of energy, corresponding to the absorption of light with a wavelength of 150–250 nm. $n \rightarrow \sigma^*$ transitions exhibit low absorptivity ($\varepsilon <$ 300), with weak absorption. Hence, a majority of these saturated hydrocarbons, such as alcohols and ethers, can also be used as solvents.

3. $\pi \rightarrow \pi^*$ transition

    In unsaturated hydrocarbons, such as conjugated alkenes and aromatic hydrocarbons, $\pi \rightarrow \pi^*$ transitions are possible. Isolated double bonds exhibit a main peak at a wavelength of less than 200 nm. Typically, conjugated double bonds absorb light in the UV region, with extremely high absorptivity ε as well as strong peaks.

4. $n \rightarrow \pi^*$ transition

    When unshared pairs of electrons and double bonds are present, $n \rightarrow \pi^*$ transitions are possible. This transition does not need significant amounts of energy; hence, the peak appears at wavelength of greater than 200 nm, albeit absorptivity is extremely low, typically 10–100; as a result, weak peaks are observed.

From the above discussion, electronic transitions vary with chemical structures. Some functional groups may exhibit more than one transition. For example, C=O possibly exhibits $\pi \rightarrow \pi^*$, $n \rightarrow \pi^*$, and $n \rightarrow \sigma^*$ transitions, correspond to the double bond, and the O atom has an unshared pair of electrons. For UV–Vis spectra of polymer materials, $\pi \rightarrow \pi^*$ and $n \rightarrow \pi^*$ transitions are mostly concerned as the absorptivities of other transitions are too marginal to be characteristic. In addition to the above four transitions, two more special transitions are observed—d–d transition and charge-transfer transition—which are typically observed for inorganic compounds. These transitions are not introduced in this chapter.

4.2.2 Basic Concepts

1. Chromophore: Functional groups exhibiting $\pi \rightarrow \pi^*$ and $n \rightarrow \pi^*$ transitions lead to clear peaks in the UV–Vis spectra, referred to as chromophores. Chromophores are functional groups with double bonds, e.g., C=C, conjugated double bonds, aromatic rings, C=O, C=S, and N=N, as well as functional groups such as -NO$_2$, -COOH, and -CONH$_2$.
2. Auxochrome: Some functional groups have an unshared pair of electrons. Although they are not chromophores, when they are connected to chromophores, they can shift the wavelength and increase absorptivity. Hence, these groups are referred to as auxochromes, e.g., -NH$_2$, -OR, -SH, -OH, and -Cl.
3. Red- and blue-shifts: The phenomenon involving a shift of peak wavelength to a long wavelength is known as the red-shift, while that to a short wavelength is known as the blue-shift (or violet shift).

4.2.3 UV–Vis Spectrometer

A UV–Vis spectrometer is employed for recording the UV–Vis spectra of samples, i.e., transmittance or absorbance of samples against wavelength. From the principle, a UV–Vis spectrometer comprises five parts:

■ Light Source
  The light source supplies UV light from a xenon lamp (190–400 nm) and visible light from a tungsten or halogen lamp (350–2500 nm). The most widely used region for commercial UV–Vis spectrometers is 200–800 nm.
■ Monochromator
  The monochromator, typically gratings, resolves a continuous spectrum into monochromatic light.
■ Sample Compartment
  A sample solution is frequently utilized for UV–Vis measurement. The solvent should exhibit good solubility for the sample but should not react with it; in addition, it should not exhibit absorbance and should be transparent in the employed wavelength range. Solvent polarity significantly affects a

UV–vis spectrum. For example, polar solvents may shift $n \rightarrow \pi^*$ transitions to short wavelengths and $\pi \rightarrow \pi^*$ transitions to long wavelengths. Solvent pH also significantly affects the spectrum.

For a liquid sample, a quartz cell is frequently used as the sample cell because it is transparent in the general wavelength range. A solid sample (e.g., a film) can be inserted in the rack and directly measured.

■ Detector

The detector is employed to determine the transmission from a sample irradiated with light with a wide wavelength range. Typically, a photomultiplier converts light signals into electrical signals.

■ Data Collection and Processing System

4.2.4 Application of UV–Vis Spectrometry

In organics, functional groups exhibit specific electronic transitions by absorbing the light of different energies. Hence, these groups exhibit characteristic absorbance at specific wavelengths. This feature is crucial for the qualitative analysis of samples. However, a significant number of chromophores exhibit peaks outside the typical determination range of a UV–Vis spectrometer (200–800 nm), or they exhibit extremely small absorptivities; hence, UV–Vis spectrometry is not suitable for qualitative analysis. Table 4.1 lists the characteristic wavelengths and molar absorptivities of common chromophores in polymers [2].

Table 4.1 Characteristic Wavelengths and Molar Absorptivities of Some Chromophores in Polymers

Chromophore	λ_{max}/nm	$\varepsilon_{max}/L/mol \cdot cm$
C=C	175	14000
	185	8000
C≡C	175	10000
	195	2000
	223	150
C=O	160	18000
	185	5000
	280	15
C=C–C=C	217	20000
⬡	184	60000
	200	4400
	255	204

Although UV–Vis spectrometry is not suitable for qualitative analysis, it exhibits high sensitivity to conjugated and aromatic structures (corresponding to a high molar absorptivity, typically greater than 10^4), rendering it suitable for the quantitative analysis of functional groups. Alternatively, it can be utilized for the selective detection of compounds containing conjugated or aromatic structures.

4.3 Infrared Spectroscopy

IR spectroscopy involves the study of the interactions between molecules and IR light. When a sample is irradiated with IR light, light of a certain wavelength is absorbed to stimulate vibrational and rotational energy transitions; hence, the light intensities of this wavelength decrease. Various molecular structures exhibit their own energy differences, corresponding to different IR wavelengths. Hence, IR spectroscopy is suitable for investigating molecular structures.

An IR spectrum is obtained by plotting transmittance or absorbance versus the wavelength of IR light. Currently, for convenience, instead of wavelength, wavenumber is frequently employed as the horizontal axis. Wavelength and wavenumber are interconvertible according to Equation 4.4:

$$\bar{v}\ (cm^{-1})\lambda(\mu m) = 10^4 \tag{4.4}$$

IR light covers a wide spectral range, including the near-IR region (12800–4000 cm^{-1}), middle-IR region (4000–400 cm^{-1}), and far-IR region (400–10 cm^{-1}) [3]. As almost all organics and a majority of the inorganics exhibit absorbance in the middle-IR region, this region is mainly discussed in this chapter.

4.3.1 Principle of Infrared Spectroscopy

Let us consider the simplest case of a diatomic molecule, which comprises two atoms connected by a bond; these atoms can be regarded as two rigid balls with masses m_1 and m_2, respectively, and the bond can be regarded as a weightless spring (Figure 4.3). d represents the equilibrium bond length, and x is the vibrational amplitude. Stretching vibrations can be regarded as simple harmonic vibrations and follow Hooke's law. Hence, the vibrational frequency v is calculated according to Equation 4.5:

$$v = \frac{1}{2\pi}\sqrt{\frac{k}{\mu}} \tag{4.5}$$

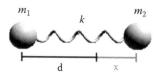

Figure 4.3 **Simple harmonic vibration model of a diatomic molecule.**

Here, ν is the frequency, Hz; k is the force constant of the bond, 10^{-5}N/cm; and μ is the reduced mass, g:

$$\mu = \frac{m_1 m_2}{m_1 + m_2} \frac{1}{N}$$

where N is the Avogadro's constant.

According to quantum mechanics, the vibrational energy of a molecule is not continuous but quantized. When a molecule absorbs energy and jumps to a high-energy level, the energy absorbed is equal to the energy difference ΔE of these two levels:

$$\Delta E = nh\nu = \frac{nh}{2\pi} \sqrt{\frac{k}{\mu}} \tag{4.6}$$

In a majority of cases, a molecule remains in the ground state. When it absorbs IR light and jumps to the first excited state (n = 1), fundamental absorption occurs. In addition, it is possible to jump to the second excited state (n = 2), and multiple-frequency absorption occurs. Because the possibility of the latter is quite low, multiple-frequency absorption is significantly weaker than fundamental absorption. In addition, multiple frequency is not two times that of fundamental frequency, but slightly lower, because the energy-level differences in a molecule are not the same.

■ Infrared Selection Rule

Not all vibrations give rise to IR absorption. Only those vibrations that result in the change of the dipole moment in a molecule can give rise to IR absorption; these vibrations are called IR-active vibrations. In contrast, vibrations resulting in no change of the dipole moment are IR-inactive vibrations, which do not give rise to IR absorption. For example, HCl exhibits IR absorption, while molecules such as N_2, O_2, and H_2 do not exhibit IR absorption.

Absorbance (peak intensity) is related not only to the number of molecules but also to the rate of change in the dipole moment. The greater the rate

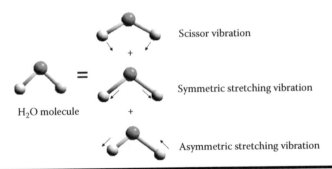

Figure 4.4 Vibrational modes in a H₂O molecule. (From http://www.columbia .edu/itc/chemistry/chem-c1403/ir_tutor/IRTUTOR.htm.)

of change in the dipole moment, the higher the absorbance. Hence, polar functional groups, such as $C=O$ and $-OH$, exhibit extremely strong IR absorption.

- Vibrational Modes in Molecules

Molecules exhibit two main vibrational modes: stretching and bending vibrations. Figure 4.4 shows the vibrational modes in a H_2O molecule [4]. Stretching vibrations include symmetric and asymmetric stretching vibrations, which result in the change of bond length. Bending vibrations include scissor vibrations, in-plane and out-of-plane bending vibrations, and swing vibrations, which result in the change of bond angle.

Nonlinear molecules with n atoms exhibit $3n-6$ vibrational modes. A linear molecule with n atoms exhibits $3n-5$ vibrational modes. Each vibration has its own energy levels and energy-level difference, corresponding to various vibrational frequencies and IR absorption bands. In multiatom molecules, some IR-active vibrations are equivalent, such as the in-plane and out-of-plane bending vibrations of CO_2 (Figure 4.5). Equivalent vibrations exhibit the same vibrational frequency; thus, only one absorption band is observed. This phenomenon is known as vibration degeneration. In addition to active vibrations, some inactive vibrations that do not generate the absorption bands that are present. Hence, the actual number of absorption bands is often less than those of vibrational modes.

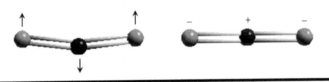

Figure 4.5 Two degenerate bending vibrations in a CO₂ molecule.

Actual molecules comprise several atoms; hence, the total number of vibrational modes is high. For brevity and convenience, functional groups are typically investigated as compared to bonds.

4.3.2 Infrared Spectrometer and Sampling

4.3.2.1 Fourier Transform Infrared Spectrometer

An IR spectrometer measures the transmittance of a sample by the absorption of IR light of different wavelengths. By principle, an IR spectrometer, similar to a UV–Vis spectrometer, contains an IR light source, a monochromator, a sample compartment, and a detector. Thus, in a traditional IR spectrometer, a grating serves as the monochromator. In this type of an IR spectrometer, the absorbance of a sample is individually measured within a narrow IR range. Hence, spectrum collection is time-consuming. An increase in the scanning speed can only be achieved at the cost of sacrificing precision. Alternatively, the improvement of precision implies that scanning speed should be decreased. Since rapid scanning and high precision cannot be simultaneously achieved, the measurement is considered to possibly be affected by environmental changes. In addition, a traditional IR spectrometer cannot meet the requirements for rapid analysis, which is frequently required for the monitoring of chemical reactions, or to be combined with gas chromatography.

A Fourier transform IR (FTIR) spectrometer, developed at the end of 1960s, constitutes a revolutionary improvement to the traditional IR spectrometer. FTIR uses Michelson's interferometer as the core part (Figure 4.6). IR light from the source reaches the beam splitter. One part of the light is reflected to the mobile mirror M_1 and then is reflected by M_1 to the beam splitter. Another part of light passes through the beam splitter to the stationary mirror M_2 and then is reflected by M_2.

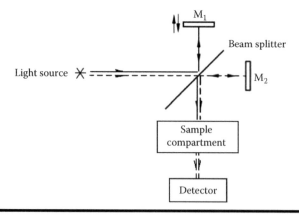

Figure 4.6   Schematic of a Fourier transform infrared spectrometer.

Interference light between the light reflected from M_1 passing through the beam splitter, and the light reflected from M_2 and the beam splitter, passes through the sample compartment to reach the detector. As M_1 continuously moves, the intensity of the interference light periodically changes with the optical path difference.

The source emits IR light with various wavelengths, which can be regarded as the addition of various monochromatic lights. Hence, the above-mentioned interference light is the addition of the interference light with these monochromatic lights (Figure 4.7). Intensity is described by Equation 4.7:

$$I(x) = \int_{-\infty}^{+\infty} B(v) \cos(2\pi v x)\, dv \qquad (4.7)$$

Here, $I(x)$ is the intensity of the interference light, which is a function of the optical path difference x. $B(v)$ is the intensity of the incident light, which is a function of frequency v.

By detecting the interference light intensity, an interferogram is obtained. By the inverse Fourier transform of the interferogram, as per Equation 4.8, an IR spectrum is obtained as a plot of transmittance/absorbance vs. wavelength, and hence the name *Fourier transform IR spectroscopy*.

$$B(v) = \int_{-\infty}^{+\infty} I(x) \cos(2\pi v x)\, dx \qquad (4.8)$$

FTIR can simultaneously measure the absorbance of a sample in a wide range of IR light, and the time needed is only approximately one thousandth of that required by the traditional grating IR spectrometer. By FTIR, rapid and precise measurement in a wide range can be achieved with high resolution. Hence, FTIR

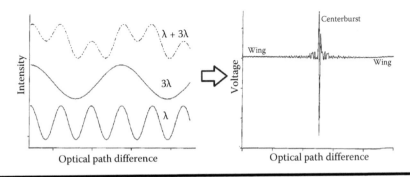

Figure 4.7   Addition of monochromatic lights to interference light.

is rapidly developing to the point at which it will nearly replace traditional IR. An FTIR spectrometer comprises the following parts:

- Infrared Light Source

  Two IR light sources exist: globar source (7800–50 cm^{-1}) and ceramic source (7800–50 cm^{-1}). The former requires water cooling during measurement; hence, it is generally used for high-resolution FTIR. The latter can be cooled by air.
- Michelson's Interferometer

  Michelson's interferometer is the core part of an FTIR spectrometer, comprising a beam splitter, a mobile mirror, and a stationary mirror.
- Sample Compartment

  The sample compartment is suitable for the measurement of solids, liquids, and gases. In addition, various accessories are available for different samples.
- Detector

  Two mid-IR detectors are employed: deuterated triglycine sulfate (DTGS) and mercury–cadmium–telluride (MCT) detectors, respectively. The former operates at room temperature, and the latter operates with liquid nitrogen. Nevertheless, the MCT detector exhibits a rapid response and high sensitivity.
- Data Collection and Processing

4.3.2.2 Sampling

Nearly all organics and a majority of inorganics (including solids, liquids, and gases), can be measured by FTIR. According to the sample nature, morphology, and analytical purpose, different sampling techniques can be employed to obtain transmission spectra.

- Solid

  Two sampling techniques are available for solids. The first technique, which is most widely utilized, is suitable for powders or brittle solids; these solids can be easily ground to fine powders. First, a powder is evenly diluted in an inert matrix powder (typically, potassium bromide, KBr) in a weight ratio of approximately 1:100. Second, the mixture is added into a mold and pressed into a small 1 mm thick pellet. Then, the pellet is placed in the sample holder, and the transmission spectrum is collected.

  The second sampling technique is suitable for polymer films that can be measured directly. If a polymer is soluble in a volatile solvent, a thin film can be prepared by solvent removal. Drops of the sample solution can be dripped on a KBr pellet and dried to obtain a thin film. When thermoplastic polymers are heated to high temperatures, these polymers melt; hence, it is convenient to produce thin films by hot-pressing the melt. The thickness of thin films is generally less than 20 μm.

Cured thermoset resins are not soluble and do not melt; hence, these resins are typically difficult to handle. Fine powders can be filed from cured resins using a file or sandpaper. The fine powders can then be sampled by the KBr pellet technique.

Lightly cross-linked samples (typically, rubber) are not soluble, but can swell with the addition of some solvents. The swollen samples can be ground together with KBr powders for even mixing, and then the solvent is removed, preparing the pellet for measurement.

Fiber samples can be obtained by cutting fibers into small pieces and mixing them with the KBr powders to prepare a pellet.

Various sampling techniques are available for a sample, somehow resulting in different IR spectra. Figure 4.8 shows three spectra of PMMA, sampled by three methods. The IR spectrum of the sample prepared by the KBr method exhibits poor quality, with a baseline shift and clear noise, caused by the fact that PMMA is rigid and can only be ground into small particles, not fine powder. These small particles scatter IR light, resulting in the baseline shift. In addition, the characteristic peaks are not clearly distinguished. In contrast, films produced by hot pressing and solution casting exhibit perfect spectra. Notably, the solvent may not be completely removed. The remaining solvent in the film shows extra peaks. For example, the small peak at 771 cm^{-1} observed in Figure 4.8 corresponds to chloroform.

Sampling methods may not only affect the spectrum quality of a polymer material, but also change its morphology, i.e., the morphology of a crystalline polymer, thereby significantly changing the spectrum. As shown in Figure 4.9, polyvinylidene chloride (PVDC) is an amorphous powder, while PVDC

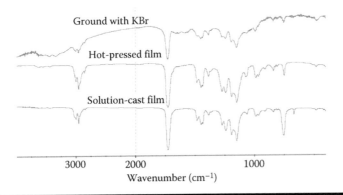

Figure 4.8   IR spectra of PMMA sampled by different methods.

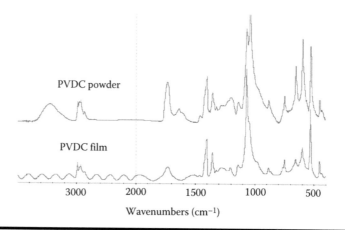

Figure 4.9 IR spectra of amorphous PVDC and crystallized PVDC.

films are semicrystalline in nature. Obvious differences are observed between the two spectra of PVDC.

■ Liquid

A liquid sample can be poured into a commercial liquid cell supplied by the manufacturer. Basically, it can also be measured by sandwiching a drop of liquid between two KBr pellets. In this case, a very thin liquid film is obtained for measuring transmittance. This method is suitable for high-viscosity, low-volatility liquids.

■ Gas

The transmission spectrum of a gaseous sample is often collected by filling the sample into a commercial gas cell.

4.3.2.3 Accessory

In addition to the above-mentioned conventional sampling methods, various convenient accessories are available for special samples.

■ Attenuated Total Reflection (ATR)

ATR is one of the most frequently used accessories in FTIR; it detects the surfaces of fibers, textiles, papers, coatings, rubbers, and elastomers. ATR-FTIR is a nondestructive method. The result does not change with sample depth.

Figure 4.10 shows the principle of ATR. First, the sample is placed on the ATR crystal surface. As the refractive index of the crystal is greater than that of the sample, the total reflection of IR light occurs between the inner crystal surfaces when the incident angle is in the appropriate range. Simultaneously, the stationary wave at each reflection point passes through the sample. A part

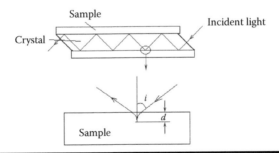

Figure 4.10   Principle of the attenuated total reflection accessory.

of the energy is absorbed by the sample; hence, the reflected light energy is attenuated and this method is called ATR.

The depth at which the amplitude of the stationary wave attenuates to 1/e of the original is known as the penetration depth d:

$$d = \frac{\lambda}{2\pi n_1 \left[\sin^2 i - \left(n_2 / n_1 \right)^2 \right]^{\frac{1}{2}}} \qquad (4.9)$$

Here, λ is the incident light wavelength; n_1 is the refractive index of the crystal; i is the incident angle at the interface between the crystal and sample; and n_2 is the refractive index of the sample.

From Equation 4.9, d depends on the wavelength and incident angle of IR light, as well as the refractive indices of the crystal and sample. In the high- and low-wavenumber regions, d differs by approximately 10 times. Hence, the spectrum must be corrected after ATR measurement to the same d value for obtaining a "common" spectrum, which is comparable to the conventional transmission spectrum (Figure 4.11).

Several ATR accessories are available, e.g., multireflection ATR, single-reflection ATR, horizontal ATR, and variable-angle ATR. Samples must be in intimate contact with the crystal surface during measurement; hence, ATR measurement is more suitable for soft matter. For powders, films, and hard solid samples, poor contact results in a low signal-to-noise ratio, thereby resulting in poor spectrum quality.

■ Infrared Microscope

An IR microscope combines the magnification characteristic of microscopy and structure analysis characteristic of IR spectroscopy; hence, it is extremely suitable for the IR analysis of very small samples or microdomains. An IR microscope can not only see what the sample looks like but also determine its composition. Hence, it is widely utilized.

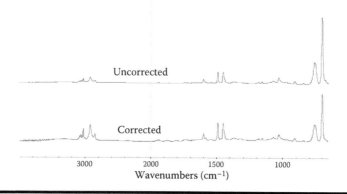

Figure 4.11   ATR spectrum before and after correction.

An IR microscope focuses IR light on a very small region (minimal diameter of 5 μm) for directly analyzing the target region of interest without the need for separation and sampling. An IR microscope does not change the original structure and morphology characteristics, therefore the samples cannot be easily contaminated.

The IR microscope is composed of a light source for imaging, an aperture, a set of objective lens, a condenser, a set of eye lens with a charge-coupled device, a sample stage, and a detector. IR light is introduced from the main optical bench. Two measurement modes are available: transmission and reflection modes. Figure 4.12 shows the optical principle of an IR microscope. In the transmission mode, IR light enters from the bottom and

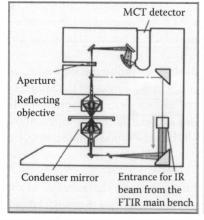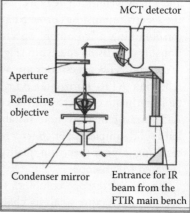

Figure 4.12   Optical principle of an infrared microscope. (Left) transmission mode; (right) reflection mode.

it is focused by the condenser onto the sample. Then, the transmitted light is refocused by the objective lens and enters the detector through the aperture. In reflection mode, the IR light enters from the top and is focused by the objective lens onto the sample surface. The reflected light then enters the detector through the aperture.

The IR microscope is a powerful tool for detecting microdomains. For example, a multilayer packaging film can be analyzed layer by layer using the IR microscope (Figure 4.13). From the spectrum of each layer, the first layer is composed of polyethylene terephthalate (PET); the second, sixth, and seventh layers are composed of polyethylene (PE); the third layer is composed of nylon; and the fourth and fifth layers are composed of ethylene–vinyl acetate copolymer (EVA).

In addition, the IR microscope can measure the spectrum along a specified direction or in a region via mapping to obtain information related to chemical structure and component distribution along the direction or in the specific range. For example, the 3D spectrum from the surface to the interior in an aged PE product was recorded to understand the oxidation profile along the thickness (Figure 4.14).

▪ Diffuse Reflectance

Diffuse reflectance is mainly utilized to obtain the IR spectra of powders. Figure 4.15 shows the schematic of its principle [5]. First, fine powders of the sample and KBr (weight ratio of approximately 1:100–1:10) are evenly mixed and added into a cup. Second, IR light is reflected by the first mirror to the focusing mirror and then to the sample powder mixture. The diffused IR light is reflected by the focusing mirror and the second mirrors to the detector. A diffuse reflectance accessory measures the intensity ratio of the diffused light to the background diffusion. The background spectrum is obtained for pure fine KBr powder in a sample cup. The sample spectrum is similar to a conventional transmission spectrum. This accessory can be

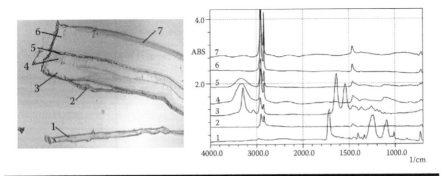

Figure 4.13 **Sectional photograph of a multilayer film (left) and spectra of these layers using the IR microscope (right).**

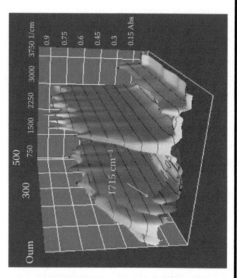

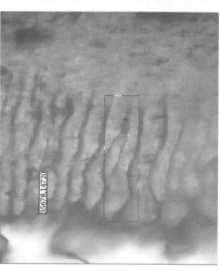

Figure 4.14 Sectional photograph of an aged PE product (left) and the 3D spectra by linear mapping (right).

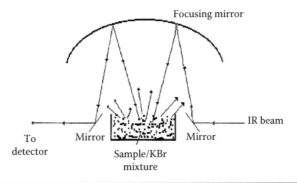

Figure 4.15 Principle of diffuse reflectance. (From Smith, B.C. *Fundamentals of Fourier Transform Infrared Spectroscopy.* Boca Raton, FL: CRC Press, 1996.)

used for room- and high-temperature measurement or for high-pressure or vacuum measurement. Hence, catalysts, as well as absorption behavior, can be conveniently studied by this technique.

▪ Specular Reflectance

A specular reflectance accessory is utilized to obtain the IR spectra of very thin films on smooth surfaces (generally metals) with high reflectivity. Figure 4.16 shows the schematic of this principle. IR is reflected from the sample film surface or the metal surface to the detector. The incident angle is equal to the reflected angle. If the thickness of the sample is on the scale of microns, the incident angle is generally 30°. If the sample is very thin (nanometer thickness), the incident angle is generally 80–85° to ensure a sufficient optical path length, which is referred to as the grazing angle reflectance.

IR spectroscopy is a universal analytical method. With abundant accessories, it is one of the most important analytical methods for obtaining structural information.

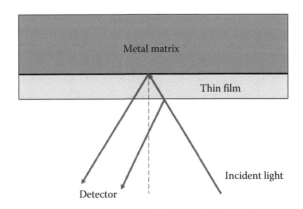

Figure 4.16 Schematic of specular reflectance.

4.3.2.4 Data Processing

In addition to the abundant accessories, powerful data processing techniques of IR spectroscopy have developed in the past decades to obtain considerable structural information. Here, some common functions are introduced:

- Baseline Correction

    When IR spectra of powders are recorded, a tilt baseline is often observed because these powders are not sufficiently fine; hence, the diffusion of powders to IR light differs with wavelength. For solving this issue, the powder must be ground to make it sufficiently fine. If this does not work, the baseline correction function in the software will help to drag the tilt baseline to the zero baseline (Figure 4.17). This function changes the intensity and not the peak position. Hence, baseline correction must be carefully carried out for quantitative analysis.
- Subtraction

    Subtraction implies the subtraction of one IR spectrum from another IR spectrum. For a mixture composed of two components (A and B), by subtracting spectrum A from the mixture spectrum, spectrum B can be obtained. As the component concentration is unknown, there must be a coefficient (subtraction factor) when subtraction is carried out:

Subtraction spectrum = mixture spectrum − spectrum A × subtraction factor

(4.10)

Notably, when there is a strong interaction between two components, the corresponding peak may shift, resulting in a negative peak when subtraction

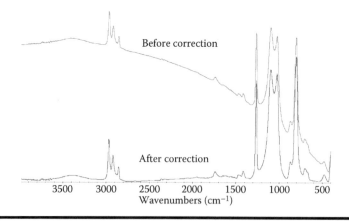

Figure 4.17   Infrared spectra before and after baseline correction.

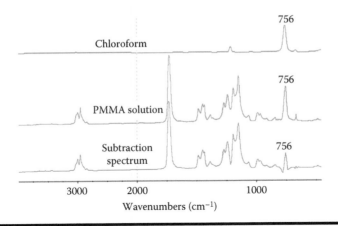

Figure 4.18   Subtraction of a PMMA solution with solvent chloroform.

is carried out. As shown in Figure 4.18, in the subtraction spectrum of a PMMA solution and solvent chloroform, a negative peak is observed around 756 cm^{-1}, which is a peak characteristic of chloroform. This is a disadvantage of the subtraction technique. Alternatively, this phenomenon can be utilized to study the interaction between two components.

■ Derivative

For overlapped bands, the derivative function helps to improve peak resolution. In a first-order derivative, the wavenumber at zero absorbance (point a–m in Figure 4.19b) corresponds to the wavenumber at the top and bottom

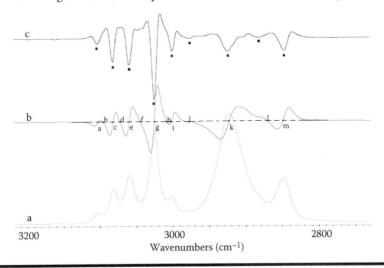

Figure 4.19   (a) Spectrum of polystyrene in the 3200–2700 cm^{-1} region. (b) First derivative of this spectrum. (c) Second derivative of this spectrum.

of the absorbance band in the original spectrum (Figure 4.19a). The derivative of the first derivative is the second-order derivative (Figure 4.19c). The bottoms exactly correspond to the wavenumber at the top of the absorbance in the original spectrum. Hence, the second-derivative spectrum is utilized for resolving the peaks of overlapped bands.

■ Curve Fitting

Curve fitting is a mathematical method for improving the resolution of a spectrum. Typically, it is performed on a section of a spectrum corresponding to a broad band caused by the overlap of several bands. An individual band can be mathematically expressed by a Gaussian function, a Lorentzian function, or the addition of both functions. Hence, the broad profile is regarded as the addition of several individual bands with different concentrations. Curve fitting involves the calculation of the number, location, width, and height, as well as shape, of these underlying peaks. The first step in curve fitting involves the second derivative by which the peak number and location can be examined. Then, a least-squares fitting algorithm is employed to obtain the optimal parameters by iteration. By curve fitting, the overlapped bands are divided, and the peak height and peak area of each band are obtained for quantitative analysis. Figure 4.20 shows the curve fitting result of ethylene-propylene-diene copolymer (EPDM) in the 3000–2750 cm^{-1} range. There are six peaks in this region, corresponding to stretching vibrations of -CH$_3$ (#1, #2 and #5), -CH$_2$ (#3 and #6), and -CH groups (#4), respectively.

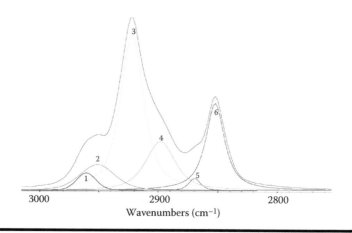

Figure 4.20 **Curve-fitting result of EPDM in the 3000–2750 cm^{-1} range.**

4.3.3 Spectral Interpretation

According to the principle of IR spectroscopy, various functional groups exhibit their own specific vibrational modes; hence, these groups absorb IR light of different wavelengths, resulting in characteristic absorptions. Only when these characteristic frequencies are known can IR spectra be interpreted and information be obtained. According to Equation 4.5, the characteristic frequency is proportional to the force constant and inversely proportional to the reduced mass. A majority of the polymers are composed of C, H, O, and N atoms, and their reduced masses are similar. Hence, the vibrational frequency mainly depends on the force constant. In this section, characteristic frequencies of typical compounds, as well as features of polymers, are introduced.

4.3.3.1 Characteristic Frequency

■ Alkane

C–C vibrations exhibit quite weak intensities; hence, C–H vibrations are mainly discussed. C–H vibrations include stretching and bending vibrations of -CH_3 and -CH_2 groups and exhibit absorptions in three regions. Stretching vibration peaks appear in the 3000–2700 cm^{-1} range. The asymmetric and symmetric stretching vibrational frequencies of -CH_3 are near 2960 cm^{-1} and 2875 cm^{-1}, while those of -CH_2 are near 2925 cm^{-1} and 2855 cm^{-1}, respectively. Bending vibration peaks appear in the 1500–1300 cm^{-1} range. The asymmetric and symmetric bending vibrational frequencies of -CH_3 are near 1460 cm^{-1} and 1375 cm^{-1}, respectively, and the bending vibrational frequency of -CH_2 is near 1465 cm^{-1}. For a carbon number of greater than 4, a characteristic in-plane wag vibration of -CH_2 is observed at approximately 720 cm^{-1}. In crystalline compounds, this band splits into two peaks at 730 cm^{-1} and 720 cm^{-1}, respectively. Figure 4.21 shows the IR spectrum of tetradecane as an example to show the three characteristic regions.

If there are branches, the stretching vibration peak of –C–H appears at approximately 2890 cm^{-1}, albeit it is often overlapped by the bands of -CH_3 and -CH_2 in this region.

■ Alkene

Alkenes are composed of C=C double bonds; hence, vibrations related to –C=C are highlighted.

The peak corresponding to the stretching vibration of -C=C, also known as the backbone vibration, is observed in the 1680–1600 cm^{-1} range. For a terminal double bond, a relatively strong peak is observed. Conversely, for an internal double bond, a relatively weak peak is observed, which disappears

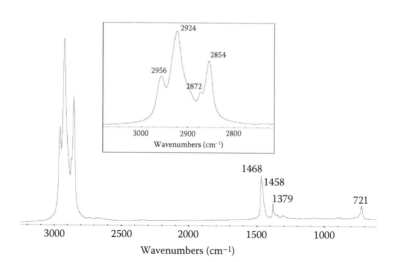

Figure 4.21 IR spectrum of tetradecane.

corresponding to the small change in the dipole moment during the stretching vibrations.

The -C–H adjacent to the double bond (-C=C–H) contributes to another characteristic of alkenes. The asymmetric stretching vibration peak is observed at 3100–3000 cm⁻¹. The symmetric vibration peak is weak and is often hidden under other peaks. The boundary between the saturated –C–H (less than 3000 cm⁻¹) and unsaturated –C–H stretching vibrations (greater than 3000 cm⁻¹) is 3000 cm⁻¹. The out-of-plane bending vibration is quite strong. For example, -C=C–CH₂ exhibits two very strong absorptions near 910 cm⁻¹ and 990 cm⁻¹, respectively; -R₁HC=CHR₂ exhibits strong absorptions near 965 cm⁻¹ (trans configuration) and 700 cm⁻¹ (cis configuration). Figure 4.22 shows the IR spectrum of n-hexene, with the labeled characteristic peaks.

■ Alkyne

Alkyne contains the -C≡C group; hence, vibrations related to -C≡C are discussed.

The stretching vibration peak of -C≡C (backbone vibration) appears at 2260–2100 cm⁻¹. Similar to the case of terminal -C=C, for a terminal -C≡C, a very sharp peak is observed, which can be easily distinguished.

The -C–H group adjacent to the -C≡C group (-C≡C–H) exhibits a strong, sharp stretching vibration peak near 3300 cm⁻¹. In addition, a strong bending vibration peak is observed at 700–610 cm⁻¹.

In addition to –C=C and -C≡C, other double and triple bonds exhibit characteristic IR absorptions in the 2500–2000 cm⁻¹ range. For example, -C≡N exhibits a characteristic peak at 2260–2240 cm⁻¹; -N=C=O exhibits a characteristic peak at 2275–2250 cm⁻¹.

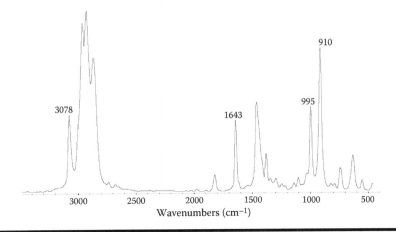

Figure 4.22   Infrared spectrum of n-hexene. (Here, only alkene-related peaks are labeled. Other characteristic peaks are not labeled to highlight features related to the double bond. In the following figures in this section, only functional group-related peaks are labeled.)

■ Aromatic Compound

Aromatic compounds are distinguished from other compounds by benzene. Hence, benzene-related vibrations are discussed.

Because of the existence of the π-conjugated structure, the backbone vibrational frequencies of the benzene ring are observed at 1600 cm⁻¹, 1500 cm⁻¹, and 1450 cm⁻¹, less than that observed for C=C. Low-intensity stretching vibration peaks of =C–H connected to the benzene ring are observed at 3100–3000 cm⁻¹. The number, wavenumber, and intensity of peaks depend on the type, number, and substituent location. Substituents also affect the fingerprint peaks in the 2000–1660 cm⁻¹ range. The out-of-plane bending vibrations of =C–H typically result in one to two strong bands in the 900–650 cm⁻¹ range, which are always the strongest peaks in a spectrum. Figure 4.23 shows the IR spectrum of toluene.

■ Alcohol and Phenol

Hydroxyl groups (-OH) are present in alcohol and phenolic compounds. In an alcohol compound, -OH groups are connected to the alkyl group. In a phenolic compound, the -OH group is connected to the benzene ring. The characteristic bands of -OH include the stretching vibration, bending vibration, wag vibration of –OH, and stretching vibration of C–OH.

The bending vibration peak (1500–1250 cm⁻¹) and wag vibration peak (750–650 cm⁻¹) easily overlap with other peaks; hence, these peaks are not characteristic. On the other hand, the stretching vibration peak is very strong. A free –OH exhibits a very sharp peak at 3650–3590 cm⁻¹, typically observed for metallic hydroxide (e.g., Mg(OH)₂ and Al(OH)₃). In an organic

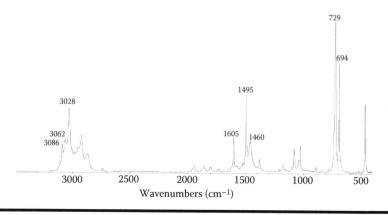

Figure 4.23   Infrared spectrum of toluene.

compound, -OH can be easily associated by hydrogen bonding; hence, the peak broadens and shifts to a lower wavenumber of 3450–3200 cm⁻¹. C–O in C–OH exhibits a strong stretching vibration band at 1300–1050 cm⁻¹. Figure 4.24 shows the IR spectra of ethanol and phenol.

Notably, the association with hydrogen bonds significantly affects all -OH vibrations, especially the stretching vibration. In addition, as –OH is present in water, strong absorptions are observed near 3300 cm⁻¹ (stretching vibration of –OH) and 1667 cm⁻¹ (bending vibration of -OH) in the IR spectrum of water. Hence, to verify if a sample contains the –OH group, the sample must be carefully dried before IR measurement to avoid interference with water (dissolved or absorbed).

■ Ether

Ether is characterized by the asymmetric stretching vibration of C–O–C, corresponding to a strong broad peak at 1150–1050 cm⁻¹ (Figure 4.25). However, several compounds, such as those containing alcohol, phenol, carboxylic acid, ester, and Si–O groups, exhibit absorptions in this region; hence, ethers are identified only in the absence of these compounds.

■ Amine

Amines are characterized by relative vibrations of the N–H group. The stretching vibration peak of N–H is observed at 3500–3300 cm⁻¹. For the free NH₂ group, two small sharp peaks are observed near 3300 cm⁻¹ (symmetric stretching vibration) and 3400 cm⁻¹ (asymmetric stretching vibration). The bending vibration peaks are observed at 1650–1500 cm⁻¹ (in-plane) and 950–600 cm⁻¹ (out-of-plane). Nevertheless, amine compounds do not exhibit strong vibration peaks, and the peaks are not easily distinguished.

■ Carbonyl Compound

The carbonyl group (C=O) is possibly one of the most important functional groups. Its stretching vibration peak is observed at 1900–1550 cm⁻¹

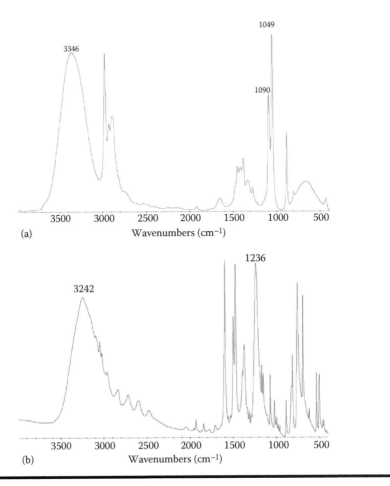

Figure 4.24 Infrared spectra of ethanol (a) and phenol (b).

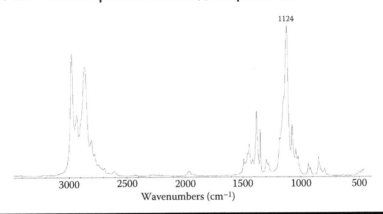

Figure 4.25 Infrared spectrum of diethyl ether.

as a strong and characteristic band nearly without interference from other bands. When the C=O group is connected to other functional groups, new functional groups, such as ketone, aldehyde, acid, ester, anhydride, amide, imide, and salts, are formed. Among these compounds, anhydride and imide groups exhibit peaks at high wavenumbers, while amide and salt exhibit peaks at low wavenumbers.

a. Ketone

In an aliphatic ketone, the C=O peak is observed at 1715 cm^{-1} (Figure 4.26a). The C=O peak in an aromatic ketone is observed at a lower wavenumber, corresponding to the conjugation effect. For example, the C=O peak in benzophenone is observed at 1665 cm^{-1}. Besides, the bending vibration peak of CH_2 connected to C=O shifts to a lower wavenumber.

b. Aldehyde

The C=O peak in aliphatic aldehydes is observed at about 1725 cm^{-1}. In an aldehyde group, the Fermi resonance between the stretching and bending vibrations of C–H results in two weak, albeit sharp, peaks at about 2820 cm^{-1} (which may be overlaid) and 2720 cm^{-1}, respectively (Figure 4.26b); these two peaks are characteristic of aldehyde.

c. Carboxylic Acid

A very strong hydrogen-bond association is observed between the C=O and –OH groups in a carboxylic acid; hence, in a liquid or solid carboxylic acid, molecules typically exist as dimers. Accordingly, characteristic peaks of C=O and –OH are simultaneously observed in the IR spectrum of a carboxylic acid. –OH exhibits a strong broad band at 3300–3000 cm^{-1}. Strong stretching vibration peaks of the C=O group are observed at 1725–1700 cm^{-1}. The stretching vibration of C–O is observed at 1350–1180 cm^{-1}. The wag vibration peak of –OH shifts to 950–900 cm^{-1}, corresponding to hydrogen-bond association (Figure 4.26c).

d. Ester

The characteristic peak of C=O in a saturated aliphatic ester is observed at 1756–1730 cm^{-1}, and the stretching vibration peak of C–O in ester is observed at 1300–1000 cm^{-1}. The asymmetric stretching vibration exhibits a peak at a relatively high wavenumber—typically greater than 1250 cm^{-1} observed for saturated aliphatic esters, while the symmetric stretching vibration peak is observed at a relatively low wavenumber (1060–1000 cm^{-1}), as shown in Figure 4.26d.

e. Amide

In an amide compound, C=O is connected to N–H; hence, the stretching vibration of C=O is affected by N–H, and the peak (often referred to as the amide-I band) is shifted to a low wavenumber (1680–1630 cm^{-1}). The coupling effect between the bending vibration of N–H and the stretching vibration of C–N results in the amide-II band at 1570–1510 cm^{-1}. C–N stretching and N–H bending vibrations result in

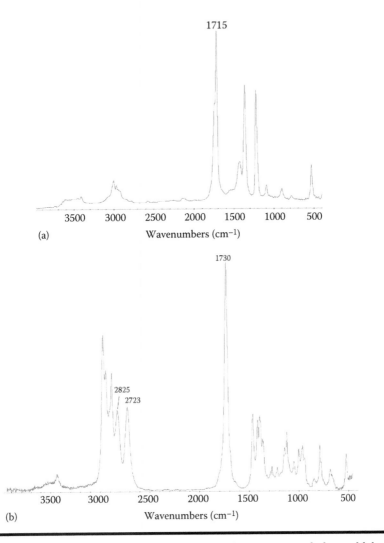

Figure 4.26 Infrared spectra of carbonyl compounds: (a) acetone, (b) butyraldehyde.
(Continued)

a weak amide-III band observed at 1335–1200 cm^{-1} and a weak amide-IV band near 700 cm^{-1} (Figure 4.26e).

  f. Anhydride

In an anhydride compound, two C=O groups are connected by an oxygen atom. The coupling of the stretching vibration of these two C=O groups results in peak splitting, affording two very strong characteristic

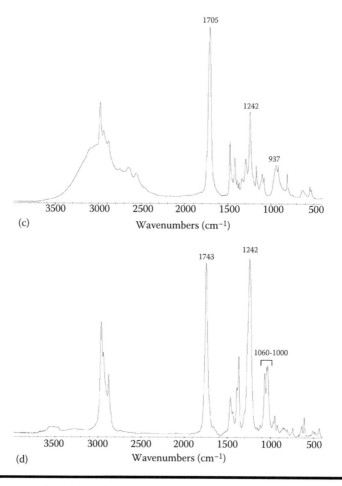

Figure 4.26 (Continued) **Infrared spectra of carbonyl compounds: (c) isobutyric acid, (d) butyl acetate.** *(Continued)*

peaks. One corresponds to the asymmetric stretching vibration (1840–1820 cm⁻¹), while the other corresponds to the symmetric stretching vibration (1770–1740 cm⁻¹). In addition, C–O exhibits a strong stretching vibration peak at 1200–1000 cm⁻¹ (Figure 4.26f).

■ Halogen-containing Compound

In a halogen-containing compound, the characteristic peak corresponds to the C–X vibration (where X = F, Cl, Br, and I). The characteristic peaks of C–F and C–Cl stretching vibration is observed at 1400–1000 and 800–600 cm⁻¹, respectively. The bending vibration peak of the C–X group is observed at 1300–1150 cm⁻¹. Figure 4.27 shows the IR spectrum of chloroform.

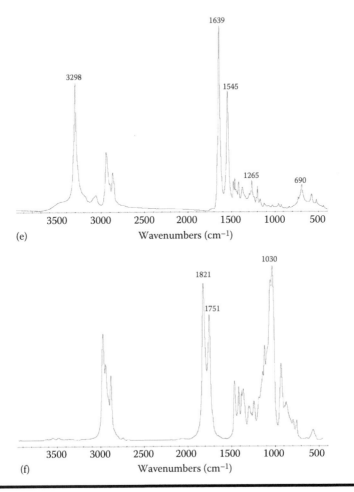

Figure 4.26 (Continued) Infrared spectra of carbonyl compounds: (e) polycaprolactam (nylon 6) and (f) butyric anhydride.

■ Silicon-containing Compound

The stretching vibration of Si–O–Si is the most characteristic peak in silicon-containing compounds, which is a very strong peak observed at 1100–1000 cm^{-1}. Other characteristic peaks include the stretching vibration peak (2300–2070 cm^{-1}) and the bending vibration peak (950–800 cm^{-1}) of Si–H, as well as the stretching vibration peak of Si–OH (910–830 cm^{-1}).

Table 4.2 summarizes the characteristic peaks observed for typical functional groups in polymers. The boundary wavenumber is 1300 cm^{-1}. Below this wavenumber, characteristic peaks from various vibrational modes of functional groups are overlapped; hence, it is typically difficult to account for each of these peaks.

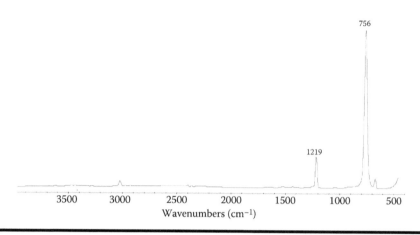

Figure 4.27 Infrared spectrum of chloroform (CHCl$_3$).

Table 4.2 Characteristic Frequencies of Functional Groups

No.	Range/cm^{-1}	Main Vibrational Modes of Functional Groups
1	4000–3000	Stretching vibration of O–H and N–H
2	3300–2700	Stretching vibration of C–H
3	2500–1900	–C≡C–, –C≡N, –C=C=C– Stretching vibration of >C=C=O and –N=C=O
4	1900–1650	Stretching vibration of >C=O, multiple frequency, and combination frequency of bending vibration of C–H in aromatics
5	1675–1500	Stretching vibration of aromatic ring, >C=C<, >C=N–
6	1500–1300	In-plane bending vibration of C–H
7	1300–1000	Stretching vibration of C–O, C–F, Si–O, and C–C
8	1000–650	Out-of-plane bending vibration of C–H; stretching vibration of C–Cl

Nevertheless, peaks in this region reflect subtle structural information, such as configuration and conformation. Hence, this region is called the fingerprint region. In contrast, characteristic peaks ranging from 4000 to 1300 cm^{-1}, referred to as the functional group region, clearly correspond to specific functional groups; hence, these peaks are often utilized for qualitative analysis.

4.3.3.2 Factors Affecting Characteristic Frequency

From the above characteristic frequencies of the various functional groups, all peaks are observed in a range, not at specific wavenumbers because when a functional group is connected to or is near other different functional groups, its chemical environment changes; as a result, its force constant and its vibrational frequency also change; this effect corresponds to both the intrinsic structure and external environment.

- Influence of the Intrinsic Structure
  - Inductive Effect

    The inductive effect corresponds to the electronegativity of substituents. When a target functional group is connected to substituents with different electronegativities, the resultant cloud density, as well as the force constant and vibrational frequency, is different. As shown in Table 4.3, when the C=O group is connected to an electronegative substituent, the stretching vibrational frequency of C=O is shifted to a high wavenumber. The higher the electronegativity, the higher the wavenumber. For bending vibrational frequency, the electronegativity of the substituent exhibits an opposite effect. As shown in Table 4.4, the higher the electronegativity, the lower the wavenumber of the bending vibrational frequency of CH_2.

  - Conjugated Effect

    In a conjugated structure, electrons move across the entire π-bond; hence, single bonds are shortened, and double bonds are extended. The

Table 4.3 Effect of Substituents on the Stretching Vibrational Frequencies of the Carbonyl Group

Compound	O‖R-C-R'	O‖R-C-Cl	O‖Cl-C-Cl	O‖F-C-F
Characteristic frequency/cm^{-1}	1715	1802	1828	1928

Table 4.4 Effect of Substituents on the Bending Vibrational Frequency of CH_2

	$-CH_2-CH_2-$	$-CH_2-CHCl-$	$-CH_2-CHF-$	$-CH_2-CCl_2-$	$-CH_2-CF_2-$
Characteristic frequency/cm^{-1}	1465	1426	1408	1400	1389

Table 4.5 Effect of Conjugation on the Stretching Vibrational Frequency of Carbonyl Group

Compound	$CH_3\text{-}\overset{O}{\underset{\|}{C}}\text{-}CH_3$	⬡-$\overset{O}{\underset{\|}{C}}$-$CH_3$	$CH_2{=}CH\text{-}\overset{O}{\underset{\|}{C}}\text{-}CH_3$	⬡-$\overset{O}{\underset{\|}{C}}$-⬡
Characteristic frequency/cm^{-1}	1715	1690	1687	1666

stretching vibrational frequency is shifted to low wavenumber, and its intensity increases. For example, when a C=O group and a benzene ring or a double bond form a conjugated structure (all the participants must be coplanar), the stretching vibrational frequency of the -C=O group decreases (Table 4.5).

– Mesomeric Effect

When an atom with an active lone pair of electrons is connected to a double or triple bond, and both these entities are coplanar, a p–π conjugation structure is formed. In this case, the stretching vibrational frequency of the double or triple bond is shifted to a low wavenumber (Table 4.6).

– Coupling Effect

When an atom is shared between two bonds or a bond is shared between two vibrations, a coupling effect is observed between the corresponding stretching or bending vibrations. If these coupling vibrations exhibit similar frequencies, a strong coupling effect is observed.

– Ring Strain Effect

In cyclic compounds, the ring strain increases with the reduction in ring size, resulting in the decrease in the stretching vibrational frequency of the internal double bonds in the ring, but the stretching vibrational frequency and intensity of the external double bond increase (Table 4.7).

■ External Environment
  – Sample Status

In a different physical state, the intermolecular interaction is different, and the vibrational frequency and intensity significantly differ. In the

Table 4.6 Effect of the Mesomeric Effect on the Stretching Vibrational Frequency of the Carbonyl Group

Compound	$R\text{-}\overset{O}{\underset{\|}{C}}\text{-}R'$	$R\text{-}\overset{O}{\underset{\|}{C}}\text{-}\ddot{N}R$	$R\text{-}\overset{O}{\underset{\|}{C}}\text{-}\ddot{O}R$	$R\text{-}\overset{O}{\underset{\|}{C}}\text{-}Cl$
Characteristic frequency/cm^{-1}	1720–1710	1700–1670	1750–1730	1815–1770

Table 4.7   Effect of Ring Strain on the Vibrational Frequency of Functional Groups

Compound	⬡	⬠	◻	⬡=O	⬠=O	◇=O
Characteristic frequency/cm^{-1}	1644	1611	1576	1715	1745	1775

gas state, the characteristic frequency is high with a sharp peak. A subtle structure is observed for the rotational spectrum of molecules, as well as the vibrational spectrum. In the liquid state, the rotational spectrum is typically not observed because of strong intermolecular interactions. Figure 4.28 shows the spectra of water in the gas and liquid states. In the solid state, the vibrational spectrum is possibly different, caused by particle size and crystal morphology (Figures 4.8 and 4.9).

– Solvent Effect

Sample molecules interact with solvent molecules in solution, which possibly results in the change of location and intensity of characteristic peaks. If a sample does not contain polar functional groups, a weak solvent effect is observed. However, if a sample contains polar functional groups, its IR spectrum is affected not only by solvent polarity, but also by the temperature and solution concentration. For example, the C=O stretching vibration peak of acetone in cyclohexane is observed at 1722 cm^{-1}, which shifts to 1710 cm^{-1} in chloroform.

– Hydrogen Bond

When an R–X–H structure in a molecule interacts with the R′–Y structure in another molecule, a hydrogen-bond-like R–X–H···Y–R′

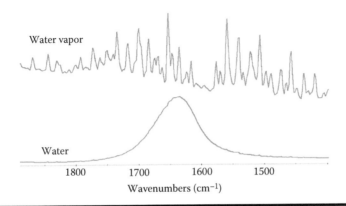

Figure 4.28   Infrared spectra of water vapor and water.

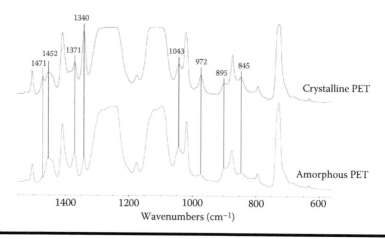

Figure 4.29 **Conformation bands in the infrared spectra of crystalline and amorphous PET.**

structure is formed, in which X is an electronegative atom, and Y is an atom with lone pair of electrons or a functional group with π-electron clouds. This hydrogen-bond association results in the broadening of the stretching vibration band of X–H, and the band is shifted to low wavenumbers. This association also results in the narrowing of the bending vibration band of X–H and shifts the band to high wavenumbers. Hydrogen bonds are often observed between acids and alcohols (phenols) or amines (Table 4.8).

In addition to intermolecular hydrogen bonds, intramolecular hydrogen bonds are also present, resulting in the decrease in the frequency and intensity of the stretching vibration.

Table 4.8   Effect of Hydrogen Bonds on the Vibrational Frequency of Functional Group (cm⁻¹)

	Absence of Hydrogen Bonds (3650–3590)			*Intermolecular Hydrogen Bond*	
Increasing acidity	Primary alcohol	~3640	Very sharp	Dimer	3550–3450 (sharp)
	Secondary alcohol	~3630		Oligomer	3400–3200 (broad)
				COOH	3000 (broad)
	Tertiary alcohol	~3615			
	Phenol	3650–3590			
	COOH	3550–3500			

4.3.3.3 Polymeric Bands

In a spectrum of a polymer, most of the characteristic peaks correspond to the repeating unit structures, which are similar to small molecules. These peaks are called monomeric bands. Some characteristic peaks are still observed, corresponding to the configuration or conformation of polymers, referred to as polymeric bands; these peaks are crucial for investigating the structure and morphology of polymers.

■ Conformation Band

Polymers in different aggregation states exhibit various conformations. For a given functional group, when polymers exhibit different conformations, the chemical environment is different; hence, vibrational frequency changes. This band is called the conformation band. For example, in PET, all of the ethyl ester groups ($-O-CH_2-CH_2-O-$) in the crystal phase exhibit trans conformations, corresponding to characteristic bands at 845 cm^{-1}, 972 cm^{-1}, 1340 cm^{-1}, and 1471 cm^{-1}. However, in the amorphous phase, some of the ethyl ester groups exhibit the trans conformation, while some exhibit the gauche conformation, corresponding to characteristic bands at 895 cm^{-1}, 1043 cm^{-1}, 1371 cm^{-1}, and 1452 cm^{-1} [6] (Figure 4.29).

■ Conformation Regularity Band

The conformation regularity band, which is related to long conformationally regular segments, corresponds to the vibrational coupling between adjacent functional groups in a polymer chain. For example, two conformationally regular bands are observed at 1167 cm^{-1} and 997 cm^{-1} in cis-PP, corresponding to the 1/3 helix structure of PP chains (Figure 4.30).

■ Crystallization Band

The crystallization band, related to a three-dimensional ordered structure of polymer chains, corresponds to the interaction between adjacent polymer chains in a crystal unit. For example, in an orthorhombic PE crystal, two

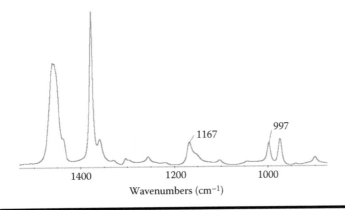

Figure 4.30   Conformationally regular bands in the infrared spectrum of cis-PP.

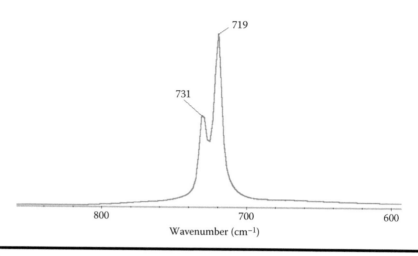

719

731

800   700   600

Wavenumber (cm^{-1})

Figure 4.31   Crystallization bands in the infrared spectrum of PE.

polymer chains are present in a crystal unit cell. The interaction between these polymer chains results in the splitting of the in-plane wag vibration band of CH_2 into two bands (Figure 4.31).

4.3.4 Applications in Polymer Materials

4.3.4.1 Qualitative Analysis

IR spectroscopy is one of the most widely used methods for the qualitative analysis of polymer materials. Positions of characteristic peaks provide information about the types of functional groups present in these materials. Peak shapes provide information regarding symmetry, isomer, and association. By qualitative analysis via IR spectroscopy, the following points need to be addressed:

1. A functional group exhibits more than one vibrational mode (e.g., asymmetric and symmetric stretching vibrations, in-plane and out-of-plane bending vibrations, and wag vibration), corresponding to more than one characteristic peak. Hence, a functional group cannot be determined by only one characteristic peak.
2. Peak position depends not only on the functional group itself but also on its chemical environment.

Some basic information with respect to the qualitative analysis of polymers is as follows (also refer to Table 4.2):

1. The strongest absorption peaks of carbonyl-containing polymers, such as polyesters, polycarboxylic acids, and polyamide, are observed in the 1800–1650 cm^{-1} range, corresponding to the stretching vibration of C=O groups.
2. The strongest absorption peaks of saturated polyolefins and polyolefins substituted by polar functional groups are observed in the 1500–1300 cm^{-1} range, corresponding to the bending vibration of the C–H group.
3. The strongest absorption peaks of polymers containing substituted benzene, unsaturated double bonds, and halogen atoms are observed in the 1000–600 cm^{-1} range.
4. The strongest absorption peaks of polyethers, polysulfones, and polyalcohols are observed in the 1300–1000 cm^{-1} range, corresponding to the stretching vibrations of C–O.

Appendix I summarizes the IR spectra of typical polymers and the identification of characteristic peaks.

The Yes–No method is often utilized for qualitative analysis. If there is no peak in the range of some functional groups, there is no functional group in the sample. For strong peaks observed in the spectrum, information provided in Table 4.2 and Section 4.3.3.1 is used as the reference to judge corresponding functional groups. Sometimes other qualitative techniques, which will be introduced in the following chapters, such as mass spectrometry and nuclear magnetic resonance, are also required for a definite conclusion.

Figure 4.32 shows the IR spectrum of an unknown sample. There is no peak at wavelengths greater than 3300 cm^{-1}, implying that hydroxyl and amine groups are absent. Characteristic peaks in the 3100–3000 cm^{-1} range correspond to the C–H

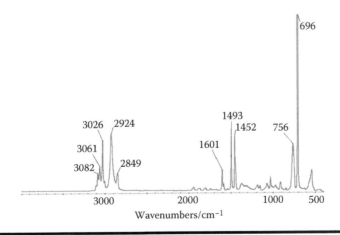

Figure 4.32 **IR spectrum of an unknown polymer.**

stretching vibrations of the benzene ring or unsaturated hydrocarbon. Peaks in the 3000–2800 cm^{-1} range correspond to the C–H stretching vibrations of saturated hydrocarbons. Peaks are absent in the 2500–1900 cm^{-1} range, indicating that nitrile and isocyanate groups are absent. Peaks are not observed in the 1900–1650 cm^{-1} range, indicating that C=O groups are absent. Peaks are observed in the 1650–1500 cm^{-1} range, indicating the presence of the benzene ring or unsaturated double bonds. Peaks are observed at 1601 cm^{-1}, 1493 cm^{-1}, and 1452 cm^{-1}, corresponding to the backbone vibration of the benzene ring. Hence, peaks observed in the range of 3100–3000 cm^{-1} are identified as C–H stretching vibrations of the benzene ring and not as unsaturated hydrocarbons. Peaks observed at 756 cm^{-1} and 696 cm^{-1} thus correspond to the C–H out-of-plane bending vibration of monosubstituted benzene. In summary, the unknown sample may be polystyrene.

Before the interpretation of the IR spectrum of a polymer, the following points should be noted:

1. The prerequisite for good spectral interpretation is to obtain a high-quality IR spectrum, which requires appropriate analytical parameters and accurate operations, such as choice of appropriate sampling technique, control of sample thickness, and elimination of the effect of background noise.
2. For the accurate interpretation of the IR spectra of unknown polymers or additives, some important information, such as sample source, properties, and application, is required for accurate interpretation.
3. An IR spectrum of a polymer is similar to that of a small molecule, which has a structure similar to the structural unit of the polymer. Nevertheless, some polymeric bands are still observed.

Currently, all manufacturers of commercial IR spectrometers supply powerful data processing software, containing search functions for verifying the similarity between an unknown spectrum and a standard library spectrum. Nevertheless, software and the standard library cannot solve all qualitative problems. If there is no similar spectrum in the library, or the unknown sample contains multiple components, the search function cannot provide a reliable result. In addition, whether the sampling methods of the unknown and standard are the same must be examined. If they are different, the peak position, shape, and intensity may differ, also resulting in the interference with the search result.

4.3.4.2 Quantitative Analysis

Peak intensity (generally represented as absorbance) is typically utilized for quantitative analysis. From the Lambert–Beer's Law (Equation 4.2), the peak absorbance (peak area or peak height) of a sample is proportional to its concentration and thickness. Besides, another factor is crucial: peak absorbance is also related to the molar absorption coefficient. Concentrations of two peaks corresponding to different functional groups cannot be directly compared from the peak absorbance values, caused by different molar absorption coefficients.

Absorbance follows the addition law. In a sample, if several components contribute to absorbance at the same position, and there is no interaction among these components, the total absorbance represents the addition of absorbance of all components (Equation 4.3).

Quantitative analysis is conducted as follows. First, choose the characteristic peak for quantitative analysis, and second, determine absorbance, and third (the last step), select an appropriate quantitative method.

The characteristic peak for quantitative analysis must correspond to the peaks that respond sensitively to the concentration change of the target component. In addition, it should be distant from other peaks to avoid interference. Furthermore, its intensity cannot be very strong or very weak. If it is very strong, it can be easily saturated. If it is very weak, a large error can be associated with the measurement. If more than two peaks are chosen, it is better to choose peaks with comparable intensities to decrease the error.

Peak absorbance can be expressed by peak height or peak area (Figure 4.33). Regardless of which parameter is used, the baseline must be subtracted to eliminate the effect of background noise. Typically, the baseline can be set as the line connecting two peak valleys adjacent to the measured peak (shown at the top of Figure 4.33) or as the baseline of the spectrum (shown at the bottom of Figure 4.33) according to the actual situation.

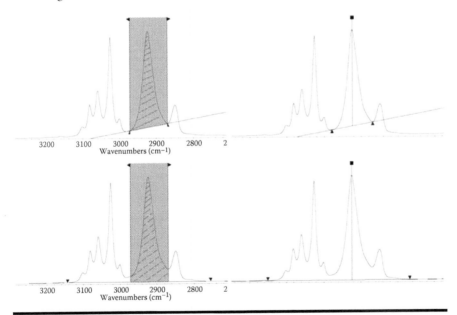

Figure 4.33 Peak absorbance by peak area (left) and peak height (right). The peak area corresponds to the area surrounded by the two boundaries, peak profile, and baseline. The peak height is the height from the top to the baseline.

Quantitative analysis by IR spectroscopy is a relative method. For determining concentration, not only the absorbance, but also sample thickness L and molar absorption coefficient ε, must be known. ε is affected by the chemical structure and environment; hence, it is typically unknown and must be determined by standard samples with known concentrations. With standard samples, a calibration curve can be determined to calculate the unknown sample concentration.

For obtaining a calibration curve, a series of standard samples with known concentrations are required, and then their absorbance values are determined. For samples with the same thickness, the calibration curve of absorbance A versus concentration c is plotted, and the unknown sample concentration can be easily calculated. If it is a linear curve, the slope is ε, which is a constant in the determined concentration range. Otherwise, the curve is nonlinear, and ε is not a constant.

In most cases, it is not typical for the samples to exhibit the same thickness. Hence, we can find a reference peak, the intensity of which does not change with concentration. By plotting the absorbance ratio of the characteristic peak to the reference peak versus concentration, the effect of sample thickness, as shown in Equation 4.11, can be eliminated. If it is difficult to find a reference peak, an internal standard with known concentration is added into the sample. By plotting the absorbance ratio of the characteristic peak to the internal standard peak versus the concentration, the effect of sample thickness can also be eliminated, as shown in Equation 4.12.

$$A_S = \varepsilon_S c_S L$$
$$A_{ref} = \varepsilon_{ref} c_{ref} L$$
$$\frac{A_S}{A_{ref}} = \frac{\varepsilon_S}{\varepsilon_{ref}} \frac{c_S}{c_{ref}} = kc_S \tag{4.11}$$

$$A_S = \varepsilon_S c_S L$$
$$A_{IS} = \varepsilon_{IS} c_{IS} L$$
$$\frac{A_S}{A_{IS}} = \frac{\varepsilon_S}{\varepsilon_{IS}} \frac{c_S}{c_{IS}} = k \frac{c_S}{c_{IS}} \tag{4.12}$$

The subscript S represents the characteristic peak of the unknown sample; *ref* represents the reference peak; and *IS* represents the internal standard peak.

For the quantitative analysis of a multicomponent sample, if the characteristic peaks overlap each other, Equation 4.13 can be utilized to build an equation set for the solutions according to the addition law. In addition, curve fitting in Section 4.3.2.4 can be utilized to obtain information corresponding to individual peaks.

For example, for determining the concentrations of methyl methacrylate (MMA) and acrylonitrile (AN) in a copolymer of MMA and AN (P[MMA-AN]), a series of standard copolymers with different MMA/AN ratios were prepared, and the nitrogen content was determined by element analysis. Figure 4.34 shows the IR spectra of the standards. The peak observed at 2241 cm^{-1} was selected to represent an AN unit, corresponding to the stretching vibration of nitrile groups. The intensities of the peaks at 1149 cm^{-1} and 987 cm^{-1} do not change with the MMA/AN ratio; hence, these peaks are selected as references. Figure 4.35 shows the relative absorbance versus the concentration of the AN unit calculated from the nitrogen content. Here, the calibration curve with 987 cm^{-1} as the reference peak exhibits

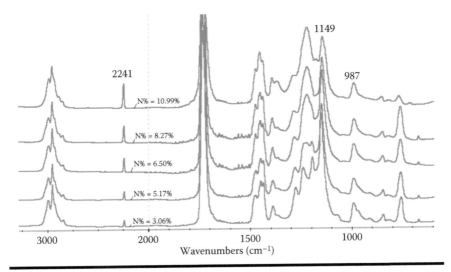

Figure 4.34 **Infrared spectra of MMA-AN copolymers with different MMA/AN ratios.**

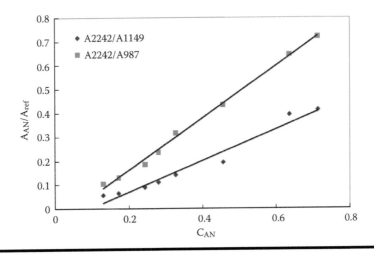

Figure 4.35 **Calibration curve of P(MMA-AN) copolymer.**

a better linear correlation as compared with that with 1149 cm⁻¹ as the reference peak. Hence, the reference peak at 987 cm⁻¹ is selected to finally conduct the quantitative analysis of unknown copolymers.

4.3.4.3 Reaction Research

An *in situ* heating accessory for IR spectroscopy is commercially available for the investigation of reaction kinetics and mechanism of polymerization and curing. In addition, polymer changes during processing and application and polymer responses to external stimulations are closely related to the changes in the molecular and aggregation structure. All these changes can also be studied by IR spectroscopy. The former *in situ* measurement is a continuous process. The measurement parameters are set according to reaction conditions. The latter needs regular sampling and measurement. Attention must be focused such that representative samples are collected and that sampling does not interfere with the reaction.

The investigation of a reaction process is divided into three steps. First, it involves the observation of the changes that occur. Second, it involves the identification of the characteristic peaks in the IR spectra reflecting changes. Third, it involves the determination of the intensity changes of these peaks during the reaction for investigating reaction kinetics. The intensity of some peaks decrease during the reaction, until they finally disappear. The reaction degree P in this case is defined as follows:

$$P = \frac{A_0 - A_t}{A_0 - A_\infty} \tag{4.13}$$

On the other hand, the intensity of other peaks increases during the reaction. The reaction degree P is defined as follows:

$$P = \frac{A_t}{A_\infty} \tag{4.14}$$

Here, A_0, A_t, and A_∞ represent the absorbance at the start, at time t, and at the end of the reaction, respectively.

Consider the polymerization of MMA as an example. During polymerization, typical changes correspond to the decrease in the C=C double bond and C–H group connected to the C=C double bond. Accordingly, unsaturated C–H is transferred to the saturated C–H group. Figure 4.36 shows the IR spectra of MMA polymerized at 75°C at different times. Obviously, the peak observed at 1639 cm⁻¹, corresponding to C=C stretching vibration, decreases with time and nearly disappears after 100 min. In addition, the intensity of the peak observed at 3106 cm⁻¹,

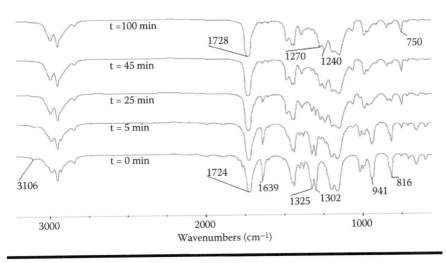

Figure 4.36   IR spectra of MMA polymerized at different times.

corresponding to the stretching vibration of C–H connected to the double bond, decreases. The intensity of the peaks observed at 941 cm^{-1} and 816 cm^{-1}, corresponding to the bending vibration of C–H connected to the double bond, also decreases. In contrast, a peak observed at 750 cm^{-1}, corresponding to the bending vibration of the formed saturated C–H, increases with time. Figure 4.37 shows the absorbance and reaction degrees of these peaks with time. Although the absolute absorbance is different for different peaks, their reaction degrees are roughly the same because they are related to the same reaction. As a result, the polymerization of MMA is an auto-acceleration reaction and is nearly completed after 40 min.

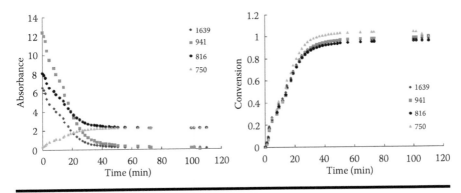

Figure 4.37   Absorbance vs. time (left) and reaction degree vs. time (right) during the polymerization of MMA.

4.3.4.4 Crystallization of Polymers

Several polymers can crystallize, but no polymer can attain 100% crystallization under normal conditions. In a polymer, the crystalline and amorphous regions coexist, where the molecular chain arrangement and chemical environment of functional groups are different. These features lead to unique bands, corresponding to the crystallization and amorphous states. The intensity of crystallization bands increases with crystallinity, contrasting to that observed for amorphous bands. Hence, these two bands can provide good understanding of the crystallization behavior and crystallization kinetics of polymers.

Table 4.9 summarizes the crystallization and amorphous bands of common polymers.

Peaks with the intensity and position independent of crystallization can be utilized as the internal standard peak; hence, sample crystallinity X_c is expressed as follows:

$$X_c = k \frac{A_i}{A_s} \qquad (4.15)$$

Table 4.9 Crystallization and Amorphous Bands of Common Polymers

Polymer	Crystallization Band/cm^{-1}	Amorphous Band/cm^{-1}
PE	1894, 731	1368, 1353, 1303
Iso-PP	1304, 1167, 998, 841	
Syndio-PP	1005, 977, 867	1230, 1199, 1131
Iso-PS	1365, 1312, 1297, 1261, 1194, 1185, 1080, 1055, 985, 920, 898	
PVC	638, 603	690, 615
PVDC	1070, 1045, 885, 752	
Polytrifluoro-chloroethylene (PTFCE)	1290, 490, 440	
PVDF	975, 794, 763, 614	657
Polyvinyl alcohol (PVA)	1144	1040, 916, 825
PET	1470, 1340, 972, 848	1450, 1370, 1045, 898
PA6	959, 928	1130
Polyamide 6,6 (PA66)	935	1140

Here, A_i and A_s represent the peak areas for the crystallization peak and internal standard, respectively. The constant k is calculated by other methods using samples with known crystallinity. For convenience, the change of A_i/A_s is investigated to represent the relative change of crystallinity. However, absolute crystallinity cannot be obtained in this case.

For a sample with unknown crystallization and amorphous peaks, a heating–cooling experiment can be conducted to identify these peaks. For example, Figure 4.38 shows the IR spectra of a crystalline co-polyolefin, which is heated from room temperature to 140°C (above the melting temperature) and then cooled down to room temperature. Peaks observed at 1167 cm⁻¹, 997 cm⁻¹, 899 cm⁻¹, 841 cm⁻¹, 808 cm⁻¹, 729 cm⁻¹, and 719 cm⁻¹ correspond to the crystallization state. Their intensities decrease with temperature and attain the lowest values at the melting state. The peak observed at 1153 cm⁻¹ corresponds to the amorphous state. Its intensity increases with temperature and attains the highest value at the melting state. Another peak observed at 972 cm⁻¹ is maintained constant during heating or cooling and can be used as the internal standard.

The necessity for a cooling procedure is doubtful, since during the heating process crystallization- and amorphous-related peaks are observed. The cooling procedure serves to remove the possible effect of chemical reactions that may also occur during heating. If a peak is related to crystallization, the change occurring during heating will recover during cooling, although the final peak intensity is different from that at the start, as the crystallinities before and after heating and cooling are different. If there is a chemical reaction, peak intensity may change or new peaks may appear. This change is irreversible and does not recover during cooling.

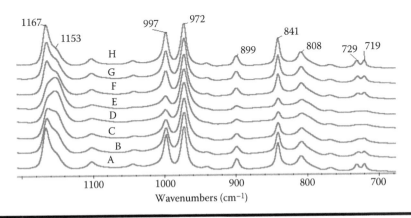

Figure 4.38 Infrared spectra of a co-polyolefin during heating and cooling A–E: heating from room temperature to 140°C; E–H: cooling from 140°C to room temperature.

4.3.4.5 Orientation of Polymers

By inserting a polarizer in the path of light, polarized IR can be recorded, which is effectively used to investigate the orientation of polymer chains.

As shown in Figure 4.39, IR light enters a polarizer and changes to polarized light. When light is irradiated onto a sample, if the vibration direction of a functional group (C=O group in Figure 4.39) is parallel to the direction of the polarized light, vibration absorption is strengthened, affording the highest peak intensity (Figure 4.39a). If the vibration direction of a functional group is perpendicular to the direction of the polarized light, the vibration absorption is weakened, affording the lowest peak intensity (Figure 4.39b). This phenomenon is called IR dichroism. For an isotropic sample, in which molecules are arranged in an unordered manner and functional groups vibrate at random directions, dichroism is not observed because of the same average absorption in all directions. On the other hand, for an oriented sample, in which molecules are arranged in an ordered manner and functional groups vibrate at specified directions, dichroism is observed because of the prevailing absorption in these directions.

For a uniaxial drawn sample, if the absorbance values corresponding to directions parallel and perpendicular to the stretching direction are represented as $A_\parallel$ and $A_\perp$, respectively, dichroism R is expressed as follows:

$$R = A_\parallel / A_\perp \tag{4.16}$$

For unordered samples, $R = 1$. If $R < 1$, the corresponding band is referred to as the perpendicular band; If $R > 1$, the corresponding band is referred to as the parallel band. Figure 4.40 shows the polarized IR spectra of PP films prepared by

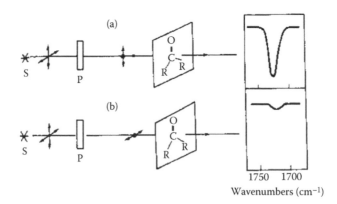

Figure 4.39 Interaction between the stretching vibration of carbonyl group and polarized light; S is the infrared light source; and P is the polarizer.

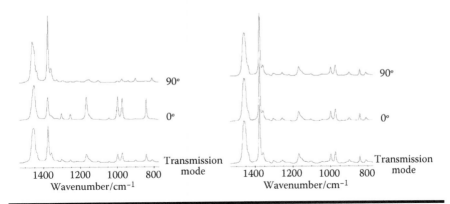

Figure 4.40 **Polarized infrared spectra of PP films prepared by uniaxial drawing (left) and solution casting (right).**

uniaxial drawing and solution casting. The angles in the figure correspond to that between polarized light and the drawing direction. For a uniaxial drawn PP film, the difference between spectra in the two directions reflects orientation. These two spectra are also different from the transmission spectrum, indicative of isotropic PP. Meanwhile, an isotropic PP film is obtained from solution casting; hence, the spectra in the two directions exhibit no difference and are the same as the transmission spectrum.

For a sample with ideal orientation, if all molecular chains are oriented along the drawing direction, and the angle between the vibration direction of a functional group and the drawing direction is α, then IR dichroism R_0 is expressed as follows:

$$R_0 = 2ctg^2\alpha \tag{4.17}$$

In fact, it is impossible for all molecular chains to be oriented along the same direction. Hence, an orientation function f is applied to represent the orientation degree of molecular chains. f representing the percentage of oriented molecular chains is expressed by Equation 4.18. That is, f completely oriented molecular chains are present, while the remaining $1-f$ represent randomly oriented molecular chains.

$$f = \frac{R-1}{R+1}\frac{2}{3\cos^2\alpha-1} = \frac{(R-1)(R_0+2)}{(R+2)(R_0-1)} \tag{4.18}$$

The angle between the polarizer and drawing direction can be changed to obtain the angle α corresponding to the maximum R_0. Then, f can be calculated according to Equation 4.18.

4.4 Raman Spectroscopy [7]

UV–Vis and IR spectroscopy are types of absorption spectroscopy, which involve the measurement of light absorption by samples when irradiated with UV–visible or IR light. On the other hand, Raman spectroscopy is a type of scattering spectroscopy, which measures the amount of light scattered by samples. The light intensity for Raman scattering is quite low, approximately $1/10^6$ of the Rayleigh scattering intensity. Hence, the development of Raman spectroscopy was slow until the introduction of the laser Raman technique.

4.4.1 Principle

When a sample is irradiated with a laser with frequency ν, a part of it is absorbed by the sample, while the majority passes through the sample and a small part is scattered. Two scattering modes exist: Rayleigh scattering and Raman scattering. In the former, the incident light exhibits elastic collisions with sample molecules, and no energy exchange occurs. The scattered light changes the direction, albeit frequency does not change. In the latter, the incident light exhibits inelastic collisions with sample molecules, and energy exchange occurs. The scattered light changes the direction and frequency. If energy is provided by light to the molecules, the energy transition from the vibration or rotation level to the higher level occurs, the energy of the scattered light decreases, and the frequency is shifted to $\nu - \dfrac{\Delta E}{h}$. The Stokes line is produced. In contrast, if energy is lost from the molecules to light, molecules undergo relaxation from a higher level, the energy of scattered light increases, and the frequency shifts to $\nu + \dfrac{\Delta E}{h}$. The anti-Stokes line is produced (Figure 4.41).

The difference between the frequencies of the Stokes line or anti-Stokes line and the incident light, $\dfrac{\Delta E}{h}$, is referred to as the Raman shift. Typically, molecules are in the ground state; hence, the intensity of the Stokes line is significantly greater than that of the anti-Stokes line, and the Stokes line is typically used to determine the Raman shift.

The Raman shift occurs in the 25–4000 cm^{-1} range. The intensity is related to the energy-level difference of molecular vibration and rotation. Hence, the Raman shift reflects the same physical nature as IR absorption, and the Raman spectrum looks the same as an IR spectrum. For the same molecular vibrational mode, the Raman shift is at the same wavenumber as the characteristic IR peak.

Not all molecular vibrations cause IR absorption. Only infrared-active vibrations lead to IR absorption. Similarly, a selection rule is applicable in Raman spectroscopy. Only vibrations involving the change in molecular polarizability α are Raman-active vibrations and exhibit Raman shifts. Molecular polarizability refers

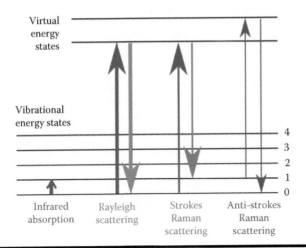

Figure 4.41 Schematic of IR absorption, Rayleigh scattering, and Raman scattering.

to the deformation degree of the electron cloud when a molecule is in an electric field E. In this case, electrons and nucleus move to opposite directions, resulting in the induced dipole moment μ_i.

$$\mu_i = \alpha E \tag{4.19}$$

Molecular vibration causes a change in molecular polarizability. The degree can be estimated by the shape difference of the electron clouds at the two ends from the equilibrium position. The greater the difference, the greater the value of α, and the stronger the Raman scattering.

Laser is a linearly polarized light. When the laser is irradiated onto molecules, different degrees of molecular polarization are observed at different directions; hence, scattered light is possibly a polarized light. Hence, the depolarization ratio ρ is used to represent molecular symmetry.

$$\rho = \frac{I_\perp}{I_\parallel} \tag{4.20}$$

Here, $I_\perp$ and $I_\parallel$ represent the intensities of peaks corresponding to absorptions in the directions parallel and perpendicular to the electric vector of the laser, respectively. For an isotropic vibration, $\rho = 0$. For an anisotropic vibration, $\rho = 3/4$. Generally, $0 < \rho < 3/4$. A small ρ implies good symmetry.

4.4.2 Laser Raman Spectrometer

A laser Raman spectrometer contains the following parts:

- Laser Source

    The laser light energy is greater than that required for vibrational energy transition, albeit less than that required for the electronic transition of molecules. Hence, lasers with wavelengths of 632.8 nm, 514.5 nm, 488.0 nm, and 1064 nm are typically used. A laser is basically linearly polarized light, which can also be utilized to measure molecular symmetry.

- Sample Compartment

    Samples include solids (films or powders) and liquids. H_2O exhibits a very weak Raman scattering signal; hence, Raman spectroscopy allows for the direct measurement of an aqueous solution without interference by H_2O.

- Detector

    The detector is employed to determine the scattered light intensity. For avoiding interference with the incident light, scattered light at 90° is typically utilized.

4.4.3 Comparison with IR Spectroscopy

Owing to the same physical nature, Raman and IR spectroscopy affords the same characteristic peaks (Raman shift corresponds to the IR wavenumber). Nevertheless, some differences in the results are still observed. IR spectroscopy is sensitive to the changes of dipole moment; hence, it is suitable for identifying polar functional groups. On the other hand, Raman spectroscopy is sensitive to the changes of the electron cloud; hence, it is suitable for identifying the molecular backbone. These two methods complement each other very well. On the other hand, they are mutually exclusive. If a functional group has a symmetric center, it is not possible for it to simultaneously exhibit IR-active and Raman-active vibrations. The asymmetric vibration of this functional group is IR active and not Raman active. The symmetric vibration of this functional group is Raman active and not IR active. If a functional group does not have a symmetric center, its vibrations are both IR and Raman active, and the IR and Raman spectra are similar. Table 4.10 shows the comparison between IR and Raman vibrations.

Table 4.10   Comparison between IR and Raman Vibrations and Characteristic Frequencies

	Vibrational Mode	*Characteristic Frequency/cm^{-1}*
Strong absorption in infrared spectroscopy*	C=O stretching	1600–1800
	C–O stretching	900–1300
	O–H stretching (hydrogen bond)	3000–3400
	Out-of-plane aromatic C–H bending	650–850
	N–H stretching (hydrogen bond)	3100–3300
	Si–O–Si asymmetric stretching	1000–1100
Strong scattering in Raman spectroscopy**	Aromatic C–H stretching	3000–3100
	C=C stretching	1600–1700
	C≡C stretching	2100–2250
	S=S stretching	1400–500
	C–S stretching	600–700
	In-plane aromatic C=C stretching	950–1050
	Aromatic ring	1500–1700
	N=N stretching	1575–1630
Strong peak in both infrared and Raman spectroscopy	Aliphatic C–H stretching	2800–3000
	C≡N stretching	2200–2300
	Si–H stretching	2100–2300
	C–X stretching (X is a halogen atom)	500–1400

* Including bending vibration of asymmetric functional groups and stretching vibrations of polar bonds.
** Including stretching vibration of symmetric functional groups and bonds (especially nonpolar and bonds).

References

1. Yang, R., X. Zhou, C. Luo, and K. Wang. *Advanced Intrumental Analysis of Polymers*, 3rd ed. Beijing: Tsinghua University Press, 2010.
2. Department of Chemistry, Wuhan University. *Instrumental Analysis*. Beijing: Higher Education Press, 2001.
3. Weng, S. *Fourier Transform Infrared Spectroscopy*, 2nd ed. Beijing: Chemical Industry Press, 2010.
4. http://www.columbia.edu/itc/chemistry/chem-c1403/ir_tutor/IRTUTOR.htm
5. Smith, B.C. *Fundamentals of Fourier Transform Infrared Spectroscopy*. Boca Raton, FL: CRC Press, 1996.
6. Shangguan, Z. and Y. Yi. Determination of crystallinity of PET by infrared spectroscopy. *Plastics Industry* 5, 1986, 54–58.
7. Wu, J. *Advanced Fourier Transform Infrared Spectroscopy: Techniques and Applications*. Beijing: Science and Technology Documentation Press, 1994.

Exercises

1. Analyze if the following polymers have UV absorption in the range of 200–400 nm and explain the reason.
   (1) Polystyrene;
   (2) Polyethylene;
   (3) Polybutadiene;
   (4) Polycarbonate
2. Calculate the wavelength range of light to excite electronic transition, vibrational transition and rotational transition, respectively.
3. How many vibration modes are there in CO_2 and H_2O, respectively? How many absorption bands appear in their IR spectra? For a PE with molecular weight of 140,000, how many vibration modes are there in it? And how many absorption bands appear in its IR spectrum? Why is the number of the vibration mode different from the number of the absorption band?
4. If the force constant of a single bond is about 500 N/m, and the force constant of a double bond or a triple bond is about two or three times of that of the single bond, calculate the stretching vibration frequencies of C-H, C=O and C≡N. Compare the results with the determined data and analyze the reason.
5. For an ATR accessory with the ZnSe crystal (refractive index is 2.42, wavelength range is 20,000–600 cm^{-1}), calculate the penetration depth of infrared light into a polymer (refractive index is 1.5) at the incident angle of 45°.
6. There are absorption bands in IR spectra of an aromatic compound (chemical formula C_7H_8O) at 3380, 3040, 2940, 1460, 690, and 740 cm^{-1}, while peaks at 1736, 2720, 1380, and 1182 cm^{-1} are absent. Analyze possible structures of this compound.
7. Compare the principles of UV–Vis, IR, and Raman spectroscopies.

Chapter 5

Nuclear Magnetic Resonance Spectroscopy

In an external magnetic field, the nucleus of an atom may exist in various spin states; thus, it can have different energies. When this nucleus is irradiated with the electromagnetic waves of an appropriate frequency, it absorbs the energy and jumps from the ground state to the excited state, producing a nuclear magnetic resonance (NMR) signal. Similar to ultraviolet–visible (UV–Vis) and infrared (IR) spectroscopy, NMR spectroscopy is also a type of absorption spectroscopy. The only difference is that in NMR spectroscopy, nuclei absorb radiofrequency (RF) energy and exhibit spin transitions; in UV–Vis spectroscopy, electrons absorb UV–Vis light and exhibit electronic transitions; while in infrared (IR) spectroscopy, molecules absorb IR light and exhibit vibrational and rotational transitions.

5.1 Basic Principle of NMR

In NMR spectroscopy, nuclei with nonzero magnetic moments are studied.

5.1.1 Magnetic Moment of Nucleus

A spinning nucleus possesses *angular momentum P*. As the nucleus has a positive charge, a magnetic field with a *magnetic moment* μ, is produced, because of spinning. The angular momentum and magnetic moment are related to each other as follows:

$$\mu = \gamma P. \tag{5.1}$$

Here, γ is the nuclear *magnetogyric ratio*, an intrinsic character of a specific nucleus. μ is expressed in units of the nuclear magneton β, where β is a constant (5.05×10^{-27} J/T). Table 5.1 summarizes the magnetic parameters of common nuclei.

According to quantum mechanics, the spin angular momentum of a nucleus is quantized, which is expressed by the *magnetic quantum number m*. *P* is numerically expressed as follows:

$$P = m\frac{h}{2\pi}. \tag{5.2}$$

Here, *h* is the Planck constant and *m* represents the corresponding stationary states or eigenstates of the nucleus, which can be *I*, (*I*–1), (*I*–2), …, –*I*. It relates to the *spin quantum number I* of the respective nucleus, where *I* can be 0, 1/2, 1, and 3/2.

Obviously, when *I* = 0, *P* = 0, indicating that the nucleus does not spin and does not produce magnetic moment; hence, an NMR signal is not observed. Thus, the first requirement for NMR is as follows: **A spinning nucleus must produce a magnetic moment**, i.e., the spin quantum number *I* ≠ 0. *I* is found to be related to the mass and atomic number (Table 5.2). For a nucleus with an even mass and even

Table 5.1 Magnetic Parameters of Common Nuclei

Nucleus	Magnetic Moment μ (β)	Magnetogyric Ratio γ (rad/(T·s))	NMR Frequency at 1.409 T (MHz)
^1H	2.7927	26.753×10^7	60.000
^{13}C	0.7022	6.723×10^7	15.086
^{19}F	2.6273	25.179×10^7	56.444
^{31}P	1.1305	10.840×10^7	24.288

Table 5.2 Relationship between the Spin Quantum Number as Well as Mass and Atomic Numbers

Mass	Atomic Number	Spin Quantum Number	NMR Signal	Nucleus
Odd number	Odd/even number	1/2	Yes	^1H, ^{13}C, ^{19}F, and ^{31}P
Odd number	Odd/even number	3/2, 5/2, …	Yes	^{17}O and ^{33}S
Even number	Odd number	1, 2, 3, …	Yes	^{14}N and ^2H(D)
Even number	Even number	0	No	^{16}O and ^{12}C

atomic number, (e.g., ^{12}C and ^{16}O, where $I = 0$), an NMR signal is not observed. A nucleus with $I = 1/2$, e.g., ^{1}H, ^{13}C, ^{19}F, and ^{31}P, is typically investigated in NMR. Among these nuclei, ^{1}H and ^{13}C are mostly discussed.

5.1.2 Nucleus in an External Magnetic Field

In the absence of an external magnetic field, the nuclear magnetic moment is randomly directed. In the presence of a static magnetic field, the nuclear magnetic moment is oriented along the magnetic field. These orientations are quantized, and the number of orientations, also called spin states, is characterized by m. Hence, the nuclear magnetic moment has $(2I+1)$ spin states, which correspond to $(2I+1)$ energy levels. The energy of each level is defined as follows:

$$E = -\gamma m \frac{h}{2\pi} B_0 \tag{5.3}$$

Here, B_0 is the magnetic field intensity.

Hence, the energy difference ΔE between two adjacent levels is as follows:

$$\Delta E = \gamma \frac{h}{2\pi} B_0 \tag{5.4}$$

For ^{1}H and ^{13}C nuclei, $I = 1/2$; hence, two spin states, i.e., $m = +1/2$ and $m = -1/2$, are present. The energy difference depends on the magnetic field intensity, as shown in Figure 5.1.

Hence, the second requirement of NMR is as follows: **In the presence of a magnetic field, the energy of a spinning nucleus can be separated into different levels.**

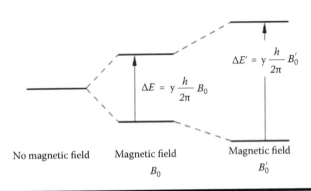

Figure 5.1   Change in the energy difference with magnetic field intensity.

5.1.3 Radiofrequency Field

In a magnetic field B_0, a nucleus has spin states with different energies. If the nucleus absorbs sufficient energy, which is equal to the energy gap, excitation occurs. A radiofrequency (RF) field supplies energy for transitions, thereby producing an NMR signal:

$$h\nu = \Delta E = \gamma \frac{h}{2\pi} B_0 \tag{5.5}$$

Here, ν is the resonance frequency of the nucleus, which is proportional to the magnetic field intensity.

Hence, the third requirement of NMR is as follows: **an RF field supplies energy for the energy transition of a spinning nucleus.**

5.1.4 Relaxation

In an equilibrium state, the number of nuclei in various spin states follows the Boltzmann distribution, and a slight excess of nuclei are present in the ground state. For nuclei with $I = 1/2$, the number ratio of nuclei in the ground state to that in the excited state is expressed as follows:

$$\frac{N_+}{N_-} = e^{\frac{\Delta E}{kT}} \approx 1 + \frac{\Delta E}{kT} = 1 + \frac{\gamma h B_0}{2\pi kT} \tag{5.6}$$

Here, N_+ and N_- represent the number of nuclei in the ground and excited states, respectively; k is the Boltzmann constant, and T is the absolute temperature.

In the case of excited nuclei that cannot return to the ground state, the number of nuclei in the ground state rapidly decrease; hence, the NMR signal rapidly decays until the transition of an atom from the ground state to the excited state ceases. Hence, some provision must be made for nuclei to return back to the ground state again to ensure a steady NMR signal, which is referred to relaxation.

Two types of relaxation exist: spin–lattice relaxation (longitudinal relaxation) and spin–spin relaxation (transverse relaxation). The former is realized by the transfer of the energy from nuclei in the excited state to the surrounding particles, which in turn results in the return of the nuclei to the ground state. The halftime of longitudinal relaxation (T_1) is 10^{-4} to 10^4 s. In a solid, this relaxation leads to the transfer of energy to the lattice. This process is difficult; hence, T_1 is long; generally, it is greater than or equal to several hours. On the other hand, in a liquid, this relaxation leads to the transfer of energy to the surrounding molecules; hence, T_1 is short (less than 1 s). Spin–spin relaxation is realized via the exchange of energy between neighboring nuclei. Nuclei in the excited state return to the ground state by transferring this energy to those in the ground state, resulting in the excitation of the nuclei in the ground state. The halftime is T_2. In a solid or viscous liquid, relaxation is easy; hence, T_2 is short. In a liquid, T_2 is approximately 1s. For a nucleus, the average time it remains in the excited state depends on which time (T_1 or T_2) is shorter.

Hence, the fourth requirement of NMR is as follows: **Nuclei in the excited state return to ground state via relaxation.**

The relaxation time affects the width of NMR peaks. According to the *uncertainty principle*:

$$\Delta E \Delta t = h \Delta v \Delta t \approx \frac{h}{2\pi}. \tag{5.7}$$

Here, Δt is the time taken by a nucleus for remaining in a specific state. That is, the shorter the relaxation time, the greater the uncertainty of ΔE. Accordingly, the greater the uncertainty of the RF frequency difference (Δv), the wider the NMR peaks. For a solid sample, the halftime of longitudinal relaxation T_1 is quite long, while that of transverse relaxation T_2 is quite short. Hence, Δt depends on T_2. A very short relaxation time results in very wide NMR peaks, and thus poor resolution. Hence, in a typical NMR measurement, a liquid sample or a sample solution is necessary to achieve an appropriate average relaxation time and good peak resolution.

5.2 Chemical Shift and Spin–Spin Coupling

5.2.1 Chemical Shift

From Equation 5.5, the resonance frequency of a nucleus is expressed as follows:

$$v = \Delta E / h = \frac{\gamma B_0}{2\pi}. \tag{5.8}$$

When the resonance condition is satisfied, a nucleus is excited, and an NMR signal is produced. From Equation 5.8, a specific nucleus, e.g., ^1H and ^{13}C, in various compounds is deduced to have the same resonance frequency in a given magnetic field. However, an induced magnetic field B_{ind} in B_0 is produced by the electron cloud surrounding it, which is opposite to B_0; hence, it weakens B_0 (Figure 5.2). The actual magnetic field intensity on the nucleus is $B_0(1 - \sigma)$, where σ is referred to as the screening (or shielding) constant; it is a measure of the chemical

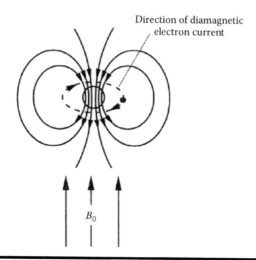

Figure 5.2   Shielding effect of the electron cloud on a nucleus.

environment around the nucleus. Hence, the actual NMR frequency is expressed as follows:

$$v' = \frac{\gamma}{2\pi} B_0 (1 - \sigma) \qquad (5.9)$$

That is, the NMR frequency changes. The frequency difference $\Delta v = v - v'$ is referred to as the chemical shift. For one nucleus in a compound, if the chemical environment (neighboring group) is different, shielding changes, resulting in different chemical shifts. Thus, the chemical shift reflects the chemical environment, which can be utilized to characterize the chemical structure of a compound. For this reason, NMR is a powerful tool for structural elucidation. Figure 5.3 shows the ^{1}H NMR spectrum of benzyl acetate [1]. The chemical environments around the protons of the methyl, methylene, and benzyl groups are different; hence, their chemical shifts are different, affording three distinct peaks.

In terms of frequency difference, the chemical shift is marginal and varies with B_0; hence, it is not practical to utilize absolute determination. Instead, a dimensionless quantity, δ, is utilized as chemical shift for convenience:

$$\delta = \frac{v_{sample} - v_{standard}}{v_0} \times 10^6 \qquad (5.10)$$

Here, $v_{sample} - v_{standard}$ represents the frequency difference between a nucleus and the standard, where v_0 is the RF frequency of the instrument (e.g., 400 MHz and

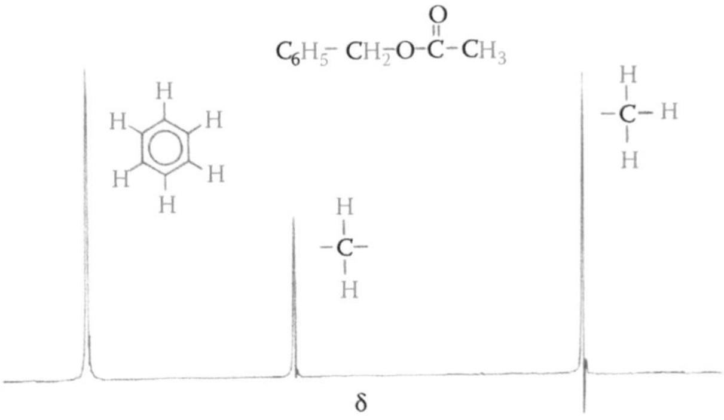

Figure 5.3 ^{1}H NMR spectrum of benzyl acetate. (Gunther, H.: *NMR Spectroscopy Basic Principles, Concepts and Applications in Chemistry*, 3rd ed. 2013. Copyright Wiley-VCH Verlag GmbH & Co. KGaA. Reproduced with permission.)

600 MHz). In NMR measurement, tetramethylsilane (TMS) is typically used as the standard. The 12 protons of TMS are strongly shielded, and only one sharp signal is observed; its chemical shift is defined as 0. Hence, δ, in ppm, is a relative value and independent of the magnetic field intensity. This relative value is convenient for the comparison of NMR spectra at different B_0 values.

In TMS, strongly shielded protons, with an extremely high electron density around them, exhibit a low resonance frequency, and hence small δ. In many other organic compounds, protons are not well shielded; hence, their chemical shifts are often greater than that of TMS, i.e., the signals are on the left of TMS. Nevertheless, this scenario does not imply that δ is always positive; it can also be negative for a proton more shielded than that in TMS.

5.2.2 Spin–Spin Coupling

By the comparison of the NMR spectra of benzyl acetate and ethyl formate in Figure 5.4, differences are observed with respect to the multiplicity of signals [1]. In the top spectrum, only singlets are observed. On the other hand, in the bottom spectrum, a singlet, triplet, and quartet are observed, each with a rather distinct intensity distribution. This observation is related to the spin interaction between protons of neighboring carbon atoms, referred to as spin–spin coupling.

Spin–spin coupling leads to the resonance peak splitting. The multiplicity of this splitting depends on the number of nuclei in the neighboring group n. For nuclei with $I = 1/2$, multiplicity equals n+1. The relative intensities for peaks of a multiplet are expressed by coefficients of a binomial expansion (Figure 5.5).

For a hydrogen nucleus, $I = 1/2$; hence, two spin states are present in the magnetic field B_0. One is aligned with B_0, and the total magnetic field is strengthened; conversely, the other is aligned opposite to B_0, and the total magnetic field is

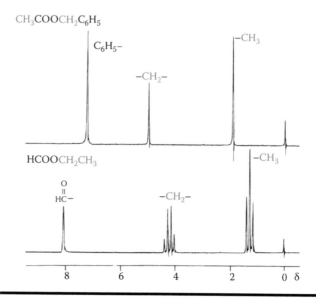

Figure 5.4 ¹H NMR spectrum of benzyl acetate (top) and ethyl formate (bottom). (Gunther, H.: *NMR Spectroscopy Basic Principles, Concepts and Applications in Chemistry*, 3rd ed. 2013. Copyright Wiley-VCH Verlag GmbH & Co. KGaA. Reproduced with permission.)

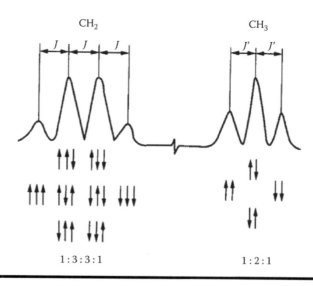

Figure 5.5 Schematic of spin–spin coupling splitting in CH_3CH_2Cl.

weakened. In $CH_3^aCH_2^bCl$, there are two hydrogen atoms in a CH_2 group; hence, the peak splitting of H^a by H^b affords a triplet with an intensity distribution of 1:2:1, corresponding to four combinations, i.e., both aligned with B_0 (↑↑), one aligned and one not aligned with B_0 (↑↓), one not aligned and one aligned with B_0 (↓↑), and both not aligned with B_0 (↓↓); among the four combinations, two combinations, ↓↑ and ↑↓, are equivalent; thus, the corresponding peaks appear at the same position (degenerate). Similarly, the peak splitting of H^b by H^a affords a quartet with an intensity distribution of 1:3:3:1, corresponding to eight combinations; among the eight combinations, six combinations are degenerate, affording two peaks.

The distance between every two split peaks in the multiplet is called the coupling constant J(Hz), which represents the strength of coupling between nuclei. J, which is typically not greater than 20 Hz, is independent of the magnetic field intensity.

If two coupled nuclei are connected by one chemical bond, ^1J is observed, e.g., ^1H–^{13}C coupling. If two coupled nuclei are separated by two chemical bonds, ^2J is observed, e.g., two protons connected to the same carbon atom. Typically, in this case, these two nuclei are chemically equivalent, and no splitting is observed. If two coupled nuclei are separated by three chemical bonds, ^3J is observed, e.g., two protons of adjacent carbon atoms. This is the most common case. If two nuclei are separated by more than three bonds, the corresponding long-range coupling is typically so weak that it is only observed in conjugated systems.

When two nuclei in a molecule are present in the same chemical environment, they exhibit the same chemical shift. These two nuclei are called chemically equivalent nuclei. For example, in CH_3CH_2Cl, the two protons in a CH_2 group, as well as the three protons in the CH_3 group, are chemically equivalent. Furthermore, if chemically equivalent nuclei exhibit the same spin–spin coupling with any other nucleus, they are referred to as magnetically equivalent nuclei. For example, in vinylidene fluoride $\underset{H}{\overset{H}{>}}C=C\underset{F}{\overset{F^a}{<}}$, two protons, as well as two fluorine nuclei, are chemically equivalent. Nevertheless, both protons and fluorine nuclei are not magnetically equivalent. Because for the fluorine nucleus a, one proton exhibits cis coupling with it, and the other proton exhibits trans coupling with it. Hence, the two protons are not magnetically equivalent, affording greater than 10 peaks in the NMR spectrum. Similarly, the two fluorine nuclei are not magnetically equivalent. In contrast, in difluoromethane $\underset{H}{\overset{H}{>}}C\underset{F}{\overset{F}{<}}$, two protons are both chemically and magnetically equivalent.

5.2.3 Spin System

A spin system is composed of spin–spin-coupled nuclei. In such a system, nuclei are coupled to each other, but these nuclei do not exhibit coupling with any other nucleus out of this system. A spin system may be either a whole molecule or a part of one molecule. For example,

$$H_3C-\bigcirc-\overset{\overset{O}{\|}}{\underset{\underset{O}{\|}}{S}}-NH-CH_2-\overset{\overset{O}{\|}}{C}-O-CH_3$$

contains three spin systems, i.e., $H_3C-\bigcirc-$, $NH-CH_2$, and $-CH_3$, respectively. The Pople notation of a spin system contains the following rules:

■ Nuclei with the same chemical shift belong to one group, denoted by a capital letter.
■ The relative chemical shift for close neighbors or neighbors that are more separated is indicated by alphabetical letters. For example, ABC indicates a strongly coupled system; AMX indicates a weakly coupled system; and ABX indicates a partially strong coupled system.
■ If nuclei in a group are also magnetically equivalent, the number of nuclei is indicated by a subscript. If nuclei in a group are not magnetically equivalent, they are distinguished by "'". For example, three such nuclei are expressed as AA'A".

Thus, the spin system $H_3C-\bigcirc-$ is named as $A_3MM'XX'$.

5.2.4 First- and Second-Order Spectra

In a spin system, the coupling strength between two nuclei groups is related to their chemical shift difference Δv (Hz) and the coupling constant J; it is evaluated by $\Delta v/J$. If $\Delta v/J > 6$, and nuclei in one group are magnetically equivalent at the same time, a first-order spectrum is obtained, with the following characteristics:

■ n magnetically equivalent nuclei are coupled with adjacent nuclei, which split the peak of the latter to $2nI+1$ peaks.
■ The relative intensities of the split peaks are expressed by the coefficients of a binomial expansion.
■ The central position of a multiplet corresponds to δ, and the distance between two peaks corresponds to J.

For example, the NMR spectrum of an AMX system is a first-order spectrum (Figure 5.6).

When a magnetically equivalent proton group is coupled with greater than one proton group, e.g., an $A_nM_pX_m$ system, the resonance signal of proton A is split to $(2pIM+1)(2mIX+1)$ peaks.

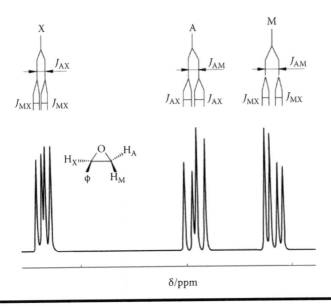

Figure 5.6   NMR spectrum of an AMX system.

If $\Delta v/J < 6$, the peak splitting does not follow the $2nI+1$ rule, and a complex intensity distribution is observed for the split peaks. In this case, a second-order spectrum is observed. For example, the spectrum of an AB system is a typical second-order spectrum (Figure 5.7). The relative intensities of the four split peaks are expressed by Equation 5.11:

$$\frac{I_2}{I_1} = \frac{I_3}{I_4} = \frac{\delta_1 - \delta_4}{\delta_2 - \delta_3}. \tag{5.11}$$

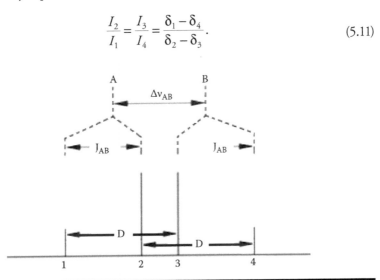

Figure 5.7   NMR spectrum of an AB system.

The chemical shift $\Delta\nu_{AB}$ is expressed as follows:

$$\Delta\nu_{AB} = \sqrt{(\delta_1 - \delta_4)(\delta_2 - \delta_3)} = \sqrt{D^2 - J_{AB}^2} \qquad (5.12)$$

An AB system is quite common, such as CH_2 connected on a ring, disubstituted ethylene, and tetrasubstituted benzene.

5.2.5 Chemical Shift as a Function of Structure

1. Electronegativity

As mentioned above, the electron cloud around a nucleus forms an induced magnetic field and exhibits a shielding effect. According to Equation 5.9, high electron density leads to strong shielding, and thus a small chemical shift. Hence, the charge density at the neighboring carbon atom becomes a determining factor for the resonance frequency of a nucleus. If a nucleus is connected to an electronegative group, its electron density decreases. The nucleus is deshielded, thereby increasing chemical shift (Table 5.3). Conversely, an electron donor increases the electron density of a nucleus, thereby decreasing the chemical shift. Typically, the electronegativity of organic functional groups is greater than that of the hydrogen atom; hence, $\delta_{CH} > \delta_{CH2} > \delta_{CH3}$. The inductive effect of neighboring groups decreases with bond separation.

2. Unsaturation Degree

If a nucleus is connected to an unsaturated group, the electron density decreases; hence, the chemical shift increases.

3. Ring-current Effect

As a special case, let us discuss the proton resonance of benzene. The chemical shift of the protons in benzene is observed at 7.3 ppm, while that in ethylene is observed at 5.23 ppm. The shielding of aromatic protons as compared to olefinic protons is reduced, caused by the circulating π-electrons surrounding the entire molecule. An aromatic molecule can be visualized as a current loop, where π-electrons freely move in a circle formed by the σ

Table 5.3 Effect of Substituent Electronegativity on the Chemical Shift δ

Compound	CH_3F	CH_3OCH_3	CH_3Cl	CH_3I	CH_3CH_3	CH_3Li
δ (ppm)	4.26	3.24	3.05	2.16	0.88	−1.95

Note: The δ values correspond to highlighted protons.

framework. If these compounds are subjected to an external magnetic field, a diamagnetic ring current is induced. The secondary field resulting from this current is aligned opposite to B_0. As a result, protons in the molecular plane and outside the ring are deshielded. Conversely, protons in the region above or below the plane of the ring are strongly shielded (Figure 5.8).

Benzene and other conjugated rings with $4n+2$ π-electrons exhibit the strong ring-current effect.

4. Solvent

In solutions, the electron density around nuclei in solute molecules are shielded or deshielded by the surrounding solvent molecules, accordingly changing the chemical shifts. For example, the resonance signals of substances dissolved in aromatic solvents are observed at small chemical shifts as compared with those of substances dissolved in aliphatic solvents. Furthermore, nuclei in solvents may also contribute to the resonance signals in NMR measurement. For avoiding complications, the use of "inert" solvents (e.g., carbon tetrachloride) or deuterated solvents without hydrogen atoms is recommended.

5. Hydrogen Bond

From experiments, the formation of both intramolecular and intermolecular hydrogen bonds leads to significant deshielding of protons and increased chemical shifts. For example, strong hydrogen bonding in carboxylic acids increases the chemical shift of protons to greater than 10 ppm. With the increase of temperature or dilution of solution, the chemical shift decreases because of the weakening of hydrogen bond. If D_2O is added instead of H_2O, resonance signals corresponding to the active protons in H_2O disappear. All of the above facts can be utilized to validate the presence of hydrogen bonds.

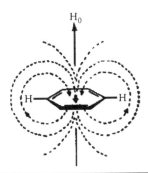

Figure 5.8   Ring-current effect in benzene.

5.3 Instrument and Sample

5.3.1 Nuclear Magnetic Resonance Spectroscopy Instrument

According to the principle of NMR, an NMR instrument comprises the following parts (Figure 5.9) [1].

- Magnet

  A magnet is required to generate an even and steady strong magnetic field for realizing energy separation. A high magnetic field and low temperature are preferred for a strong resonance signal, and thus high sensitivity and resolution; hence, NMR instruments typically utilize superconducting magnets operating at the temperature of liquid nitrogen or helium.

- RF Source/Detector

  An RF electromagnetic wave, produced by an RF transmitter, is required to supply energy for nuclear spin energy transition or resonance. Because of the shielding or deshielding of a nucleus, the resonance frequency varies in a wide range (Equation 5.9). Resonance can be realized via two methods: By varying the magnetic field intensity while simultaneously maintaining a frequency constant (field sweep) or by varying the frequency by maintaining a field constant (frequency sweep). When the resonance condition is satisfied, an NMR signal is recorded by an RF receiver and amplified for output.

- Sample Probe

  A sample probe typically contains a sample tube, a transmitter coil, and a receiver coil.

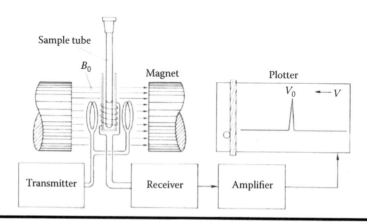

Figure 5.9 Schematic of an NMR instrument. (Gunther, H.: *NMR Spectroscopy Basic Principles, Concepts and Applications in Chemistry*, 3rd ed. 2013. Copyright Wiley-VCH Verlag GmbH & Co. KGaA. Reproduced with permission.)

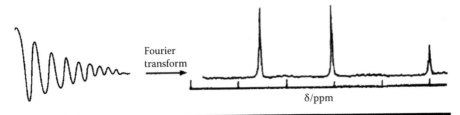

Figure 5.10   Fourier transform of the FID signal to a conventional NMR spectrum.

Such an instrument is called a continuous-wave NMR (CW-NMR). In a CW-NMR, only one type of nucleus is excited each moment, i.e., monochromatic resonance, irrespective of the field sweep or frequency sweep condition. Therefore, CW-NMR is quite time-consuming (typically greater than or equal to 250 s) as each signal needs to be separately measured.

For overcoming these disadvantages, Fourier transform NMR (FT-NMR) was developed. In a fixed magnetic field, a range of short (typically 10–50 µs), but strong, RF pulses (broad-frequency spectrum) are utilized for the excitation of nuclei (polychromatic resonance). Between two pulses, the signal fades away with time via relaxation, referred to as free-induction decay (FID). This signal, originating from all nuclei, represents an interference spectrum of all resonance frequencies. This spectrum is then Fourier transformed to a conventional NMR spectrum (Figure 5.10). In FT-NMR, a pulse, lasting for only tens of microseconds, supplies the same information as that provided from a total sweep in CW-NMR. Hence, FT-NMR is rapid and sensitive. Currently, when NMR spectroscopy is mentioned, typically it refers to FT-NMR spectroscopy.

5.3.2 *Sample*

As mentioned in Section 5.1.4, a liquid or solution is preferred for NMR measurement to obtain sharp peaks with high resolution. Typically, 5–10 mg or 10 µL of a sample for ^{1}H NMR (or 10 mg of a sample for ^{13}C-NMR) is placed in a sample tube and dissolved in 0.5 mL of a solvent with the standard TMS (5 vol%). Protons in TMS are strongly shielded and they exhibit only one sharp peak at 0 ppm. Protons in several other organic compounds are not well shielded; hence, their resonance peaks appear to the left of the TMS peak.

For avoiding the "impurity" peaks from the solvent molecule, inert solvents such as CCl_4 and carbon disulfide are used for ^{1}H NMR measurement. Typically, deuterated solvents are more often selected to dissolve various samples. However, trace amounts of protons in deuterated solvents are still present, which contribute to the small "impurity" peaks in the spectrum. Tables 5.4 and 5.5 summarize the chemical shifts of residual protons and residual carbon nuclei in commonly used deuterated solvents observed in ^{1}H NMR and ^{13}C NMR spectra, respectively.

Table 5.4 Chemical Shifts of Residual Protons in Deuterated Solvents Observed in ^{1}H NMR Spectra

Solvent	δ/ppm	Solvent	δ/ppm
CD_3COCD_3	2.05	Water in CD_3SOCD_3	3.1
CH_3COCH_3	2.07	D_2O	4.70
CD_3SOCD_3	2.50	[D_5]pyridine	6.98, 7.35, and 8.50
CH_3SOCH_3	2.50	[D_6]benzene	7.20
[D_7]dimethyl formamide	2.74 and 2.90	$CDCl_3$	7.27
Water in CD_3COCD_3	2.7	Trifluoroacetic acid	11.6

Table 5.5 Chemical Shifts of Carbon Nuclei in Deuterated Solvents Observed in ^{13}C NMR Spectra

Solvent	δ/ppm	Solvent	δ/ppm
$CDCl_3$	77.0	[D_6]benzene	128.7
CD_3COCD_3	30.2, 206.8	[D_5]pyridine	123.5, 135.5, and 149.38
CD_3OD	49.3	CD_3SOCD_3	39.7

5.4 ^{1}H NMR and ^{13}C NMR Spectroscopy

5.4.1 ^{1}H NMR Spectroscopy

In the past decades, when NMR was discussed, ^{1}H NMR was implied, especially before the development of FT-NMR, and consequently ^{13}C-NMR. ^{1}H NMR spectroscopy was first developed in the 1950s. From then, important concepts such as *chemical shift, spin–spin coupling,* and *spin systems* were further developed. ^{1}H is a stable isotope with a natural abundance as high as 99.98% and a high magnetic moment. Hence, ^{1}H NMR spectroscopy is extremely sensitive and a powerful tool for various applications.

In ^{1}H NMR spectroscopy, functional groups and their connections are characterized by chemical shifts of protons in various chemical environments. Figure 5.11 shows the ^{1}H NMR spectrum of benzyl acetate as an example [1]. From the number

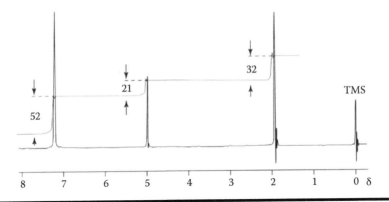

Figure 5.11 ¹H NMR spectrum of benzyl acetate. (Gunther, H.: *NMR Spectroscopy Basic Principles, Concepts and Applications in Chemistry*, 3rd ed. 2013. Copyright Wiley-VCH Verlag GmbH & Co. KGaA. Reproduced with permission.)

of peaks (groups), the number of groups of protons can be obtained. From the chemical shifts, the chemical structure (functional groups) is deduced. Moreover, from the peak splitting and coupling constant, the connection between functional groups is identified. From the step height (integration of peak area), the relative number of protons is obtained. Hence, which can be observed from Figure 5.11, three types of protons are observed, with a proton ratio of approximately 5:2:3, corresponding to protons of the benzene ring, methylene, and methyl, respectively; their corresponding chemical shifts are observed at 7.22, 5.00, and 1.93 ppm.

For reference, Figure 5.12 shows the chemical shift ranges of some important protons in organic functional groups [2]. Table 5.6 summarizes the chemical shifts of protons in a series of characteristic organics.

5.4.1.1 Empirical Expression for the Proton Chemical Shift [3–5]

Because of the contribution of several researchers who have accumulated significant data, some empirical expressions for calculating proton chemical shifts have been proposed:

■ CH₂ and CH Groups
    The chemical shifts of protons in the CH₂ and CH groups have been calculated by the following equations:

$$\delta_{-CH_2-} = 1.25 + \sum \sigma_i \tag{5.13}$$

$$\delta_{-CH-} = 1.50 + \sum \sigma_i \tag{5.14}$$

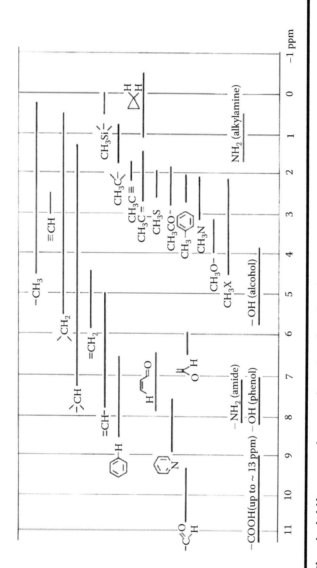

Figure 5.12 Chemical shift ranges of protons in various functional groups. (With kind permission from Springer Science+Business Media: Hatada, K., and T. Kitayama, *NMR Spectroscopy of Polymers*, 2004.)

Table 5.6 Chemical Shifts of Protons (Underlined) in Characteristic Organics

	δ/ppm		δ/ppm
$C_3\underline{H}_6$	0.22	$C\underline{H}_3CH_2OH$	1.22
$C\underline{H}_3C\underline{H}_3$	0.88	$CH_3C\underline{H}_2OH$	3.70
$C\underline{H}_2{=}C\underline{H}_2$	5.84	$CH_3CH_2O\underline{H}$	2.58
$\underline{H}C{\equiv}C\underline{H}$	2.88	$C\underline{H}_3COOH$	2.10
$C_6\underline{H}_6$	7.27	$CH_3COO\underline{H}$	8.63
$CH_2{=}CHC\underline{H}_3$	1.71	$C\underline{H}_3CHO$	2.20
$CH{\equiv}CC\underline{H}_3$	1.80	$CH_3C\underline{H}O$	9.80
$C\underline{H}_3COC\underline{H}_3$	2.17	$(C\underline{H}_3CH_2)_2O$	1.16
$C_6\underline{H}_{12}$	1.44	$(CH_3C\underline{H}_2)_2O$	3.36
$C\underline{H}_3Cl$	3.10	$C\underline{H}_3COOCH_2CH_3$	2.03
$C\underline{H}_2Cl_2$	5.30	$CH_3COOCH_2C\underline{H}_3$	1.25
$C\underline{H}Cl_3$	7.27	$CH_3COOC\underline{H}_2CH_3$	4.12
$N(C\underline{H}_3)_3$	2.12	$C_6H_5C\underline{H}_3$	2.32
$N(C\underline{H}_2CH_3)_3$	2.42	$C_6H_5C\underline{H}O$	9.96

Here, σ_i is the shielding constant for substituents summarized in Table 5.7. In addition, the chemical shifts of protons in CH_2 can be calculated according to *Shoolery's* rule:

$$\delta = 0.23 + \sum \sigma_i \tag{5.15}$$

Table 5.8 summarizes the Shoolery's effective shielding constants.
■ H–C=C
The chemical shifts of protons of a double bond are calculated as follows:

$$\delta_{C=CH} = 5.25 + Z_{geminal} + Z_{cis} + Z_{trans} \tag{5.16}$$

Here, $Z_{geminal}$, Z_{cis}, and Z_{trans} represent the effects of substituents at the same carbon atom as well as *cis* and *trans* positions of the target proton, respectively. Table 5.9 summarizes the values.

Table 5.7 Shielding Constants for Substituents

Substituent	σ_i (ppm)	Substituent	σ_i (ppm)	Substituent	σ_i (ppm)
-Cl	2.0	-NR₂	1.0	-Ph	1.3
-Br	1.9	-NH₂	1.0	-CN	1.2
-I	1.4	-SR	1.0	-R-	0.0
-OH	1.7	-CHO	1.2	-C=C	0.8
-OR	1.5	-OCOR	2.7	-C≡C	0.9
-OPh	2.3	-COR	1.20	-NO₂	3.0
-COOH	0.8	-CO₂R	0.7		

Table 5.8 Substituent Constants According to Shoolery's Rule

Substituent	σ_i (ppm)	Substituent	σ_i (ppm)	Substituent	σ_i (ppm)
-Cl	2.53	-OPh	3.23	-COR	1.70
-Br	2.33	-NR₂	1.57	-CO₂R	1.55
-OH	2.56	-CH₃	0.47	-CN	1.70
-OR	2.36	-SR	1.64	-C=C	1.32
-Ph	1.85	-OCOR	3.13	-C≡C	1.44

For example, the chemical shifts of H_a and H_b in $\underset{H_a}{\overset{C_6H_5}{}}C=C\underset{H_b}{\overset{OC_2H_5}{}}$ are calculated as follows:

H_a: $\delta = (5.25 + 1.38 + 0 - 1.21)$ ppm $= 5.42$ ppm
H_b: $\delta = (5.25 + 1.22 + 0 - 0.07)$ ppm $= 6.4$ ppm

■ Protons of the Benzene Ring
   The chemical shifts of protons of the benzene ring can be calculated as follows:

$$\delta = 7.26 + \sum Z_i \tag{5.17}$$

Here, Z_i represents the effect of substituents on the chemical shifts of protons of the benzene ring (Table 5.10).

Table 5.9   Effect of Substituents on the Chemical Shift of the Double Bond Proton

Substituent	$Z_{geminal}$	Z_{cis}	Z_{trans}	Substituent	$Z_{geminal}$	Z_{cis}	Z_{trans}
-H	0	0	0	-CN	0.27	0.75	0.55
-R	0.45	−0.22	−0.28	-OCOR	2.11	−0.35	−0.64
-R(ring)	0.69	−0.25	−0.28	-Cl	1.08	0.18	0.13
-CH$_2$O,I	0.64	−0.01	−0.02	-Br	1.07	0.45	0.55
-CH$_2$S-	0.71	−0.13	−0.22	-I	1.14	0.81	0.88
-CH$_2$F,Cl,Br	0.70	0.11	−0.04	-Ar	1.38	0.36	−0.07
-CH$_2$-N<	0.58	−0.10	−0.08	-CHO	1.02	0.95	1.17
-CH$_2$-C=O	0.69	−0.08	−0.06	-CH$_2$-Ar	1.05	−0.29	−0.32
-C=C	1.00	−0.09	−0.23	-CO$_2$H	0.97	1.41	0.71
-C=C- (conjugated)	1.24	0.02	−0.05	-CO$_2$H (conjugated)	0.80	0.98	0.32
-C=O	1.10	1.12	0.87	-C≡C-	0.47	0.38	0.12
-C=O (conjugated)	1.06	0.91	0.74	-OR (saturated R)	1.22	−1.07	−1.21
-N–R	0.80	−1.26	−1.21	-OR (conjugated R)	1.21	−0.60	−1.00
-N–R (conjugated)	1.17	−0.58	−0.99				

5.4.1.2 Elucidation of Structure from ^{1}H NMR Spectrum

^{1}H NMR spectroscopy is very powerful for structural determination. From ^{1}H NMR spectrum, the following information can be obtained:

- When a molecular formula is supplied, the unsaturation degree Ω is first calculated according to Equation 2.30. If $\Omega = 0$, a saturated molecule is expected; if $\Omega = 1$, a double bond or a ring is possibly present; if $\Omega = 4$, the presence of a benzene ring is predominantly present.
- The relative intensities of resonance signals (groups) are integrated, which indicate the types of protons and their molar ratios.

Table 5.10 Effect of Substituents on the Chemical Shift of Protons of the Benzene Ring

Substituent	Z_{ortho}	Z_{meta}	Z_{para}	Substituent	Z_{ortho}	Z_{meta}	Z_{para}
-H	0.00	0.00	0.00	-OH	−0.56	−0.12	−0.45
-CH$_3$	−0.20	−0.12	−0.22	-OCH$_3$	−0.48	−0.09	−0.44
-CH$_2$CH$_3$	−0.14	−0.06	−0.17	-OCH$_2$OH	−0.46	−0.10	−0.43
-CH(CH$_3$)$_2$	−0.15	−0.08	−0.18	-OCOCH$_3$	−0.25	0.03	−0.13
-C(CH$_3$)$_3$	0.02	−0.08	−0.21	-CONH$_2$	0.61	0.10	0.27
-CH$_2$Cl	0.00	0.00	0.00	-CHO	0.56	0.22	0.29
-CH$_2$OH	−0.07	−0.07	−0.07	-COCH$_3$	0.60	0.14	0.21
-CH=CH$_2$	0.06	−0.08	−0.10	-Cl	0.03	−0.02	−0.09
-CH=CH-Ph	0.15	−0.01	−0.16	-Br	0.18	−0.08	−0.04
-CH≡CH	0.15	−0.02	−0.01	-COOH	0.83	0.18	0.27
-Ph	0.37	0.20	0.10	-COOCH$_3$	0.71	0.11	0.21

■ Possible functional groups are proposed according to chemical shifts, and the connections between groups are also assumed on the basis of peak splitting by spin–spin coupling.
■ Possible molecular formulas are proposed and validated.

Example

From the ^{1}H NMR spectrum shown in Figure 5.13, let us determine the accurate structure.

There are five groups of signals with a molar ratio of 5:1:2:3:3 from left to right, corresponding to the presence of five protons; hence, structures (c) and (d) are not possible as only four types of protons are present in (c) and two types of protons in (d). In addition, structure (a) is not possible as the molar ratio of protons is 5:2:2:2:3. Hence, structure (b) is the correct structure. Furthermore, the triplet observed at δ = 0.9 ppm indicated that the peak is split by a CH$_2$ group. The doublet at δ = 1.2 ppm indicated that the peak is split by a CH group. Two quartets observed at δ = 1.7 ppm and δ = 2.8 ppm indicate that they are also connected to the CH$_3$ group. In conclusion, the signals are assigned as follows:

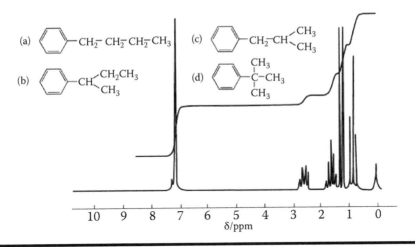

Figure 5.13 **¹H NMR spectrum of an unknown compound.**

Notably, Figure 5.13 shows a low-resolution spectrum; hence, the five protons of the benzene ring are not distinguished. In fact, the ortho, meta, and para protons are not chemically equivalent and exhibit three peaks in a high-resolution spectrum.

Currently, a majority of commercial NMR manufacturers supply software to elucidate NMR spectra and propose possible structures. However, for complex spectra, it is difficult to elucidate the chemical structure only by NMR, and other structural analysis methods are also required.

5.4.2 ¹³C NMR Spectroscopy

¹³C NMR spectroscopy involves the investigation of carbon nuclei, corresponding to the backbone in compounds. Hence, ¹³C NMR is crucial in molecular structure analysis, especially for functional groups without hydrogen atoms, such as C=O, C≡N, and quaternary carbons.

5.4.2.1 Comparison of ¹³C NMR with ¹H NMR

▪ Sensitivity

   The NMR signal intensity is proportional to $I(I+1)$ $\gamma^3 B_0^2$. From Table 5.1, the γ of a carbon nucleus is approximately 1/4 of that of protons; hence, the ¹³C NMR signal intensity is significantly less than that of ¹H NMR. In addition, the natural abundance of the ¹³C isotope is only 1.1%, while that of the ¹H isotope is as high as 99.98%. Hence, the sensitivity of ¹³C NMR spectroscopy is considerably less than that of ¹H NMR spectroscopy, only approximately 1/6000 of the latter. Because of this poor sensitivity, it is difficult to detect the resonance signal; hence, early research on NMR predominantly focuses on protons. Poor sensitivity was only resolved after the invention of FT-NMR. From then, several techniques have been developed to simultaneously improve the sensitivity and resolution, and applications of ^{13}C NMR spectroscopy have been rapidly progressing.

- Resolution

  The range of chemical shifts in ^{13}C NMR spectroscopy is approximately 0 to 300 ppm, while that in ^{1}H NMR spectroscopy is only approximately 15 ppm. Hence, ^{13}C NMR spectroscopy exhibits higher resolution.

- Topic of Investigation

  ^{13}C NMR spectroscopy involves the measurement of resonance signals corresponding to the carbon nuclei comprising the backbone of molecules; hence, information about certain functional groups, such as C=O, C≡N, and quaternary carbons, is easily obtained, which cannot be obtained using ^{1}H NMR spectroscopy.

- Spin–spin Coupling

  Because of the very low natural abundance of ^{13}C nuclei, the coupling between two ^{13}C nuclei can be neglected, which in turn makes spectral interpretation easy. Nevertheless, both protons of a carbon nucleus and protons of the adjacent carbon nucleus exhibit spin–spin coupling with the carbon nucleus, and with a large coupling constant. Hence, necessary information about C–H connections is obtained, which also renders some difficulty in spectral interpretation.

5.4.2.2 Decoupling Techniques in ^{13}C NMR Spectroscopy

The coupling between protons and ^{13}C nuclei leads to complex spectra. For simplifying and clarifying spectra, some decoupling techniques have been developed (Figure 5.14) [3].

- Broadband Decoupling

  During measurement, broadband RF radiation, including all resonance frequencies of protons, is introduced to completely eliminate ^{13}C–^{1}H coupling. Then, a ^{13}C spectrum with only singlet peaks is obtained. This is called broadband decoupling or complete decoupling. Figure 5.15 shows a broadband-decoupled spectrum of compound A, indicating the presence of 9 carbon nuclei in the molecule [3].

- Off-resonance Decoupling

  By broadband decoupling, spectra are simplified, and characteristic carbon nuclei are highlighted at the expense of missing useful information. When additional RF irradiation frequency is adjusted to hundreds to thousands of hertz away from the resonance frequency of protons, the neighboring coupling and long-range coupling are eliminated, while direct ^{13}C–^{1}H coupling is

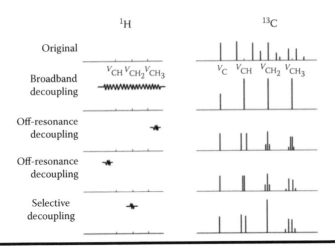

Figure 5.14 **Proton-decoupled techniques and the resulting spectra. (From Ning, Y.C.** *Structural Identification of Organic Compounds and Organic Spectroscopy,* **2nd ed. Science Press, 2000.)**

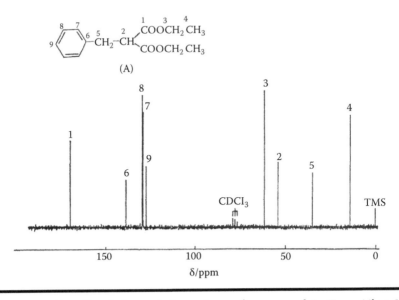

Figure 5.15 **Broadband-decoupled spectrum of compound A. (From Ning, Y.C.** *Structural Identification of Organic Compounds and Organic Spectroscopy,* **2nd ed. Science Press, 2000.)**

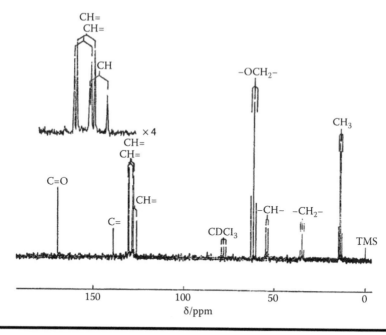

Figure 5.16 Off-resonance-decoupled spectrum of compound A. (From Ning, Y.C. *Structural Identification of Organic Compounds and Organic Spectroscopy,* 2nd ed. Science Press, 2000.)

maintained to indicate the number of protons connected to a carbon nucleus. For example, a CH_3 group exhibits a quartet, and a CH_2 group exhibits a triplet. This is called off-resonance decoupling. Figure 5.16 shows the off-resonance-decoupled spectrum of compound A [3]; a hydrogen atom is not present on #1 and #6 carbon atoms, while one hydrogen atom is present on #2, #7, #8, and #9 carbon atoms; two hydrogen atoms are present on #3 and #5 carbon atoms, and three hydrogen atoms are present on #4 carbon atom.

■ Selective Decoupling

   If the additional RF frequency is chosen to be the same as the resonance frequency of a specific proton, the coupling effect of this proton to the carbon nucleus is eliminated, and a singlet is observed. In Figure 5.17a and b, the selective decoupling of the #4 and #3 carbon atoms indicates that they are related to the CH_3 and CH_2 groups, respectively [3].

5.4.2.3 Information from ^{13}C NMR Spectra

Similar to ^1H NMR spectroscopy, ^{13}C NMR spectroscopy also provides information such as chemical shift, peak intensity, spin–spin coupling, and coupling constant for elucidating chemical structures. However, because of some reasons

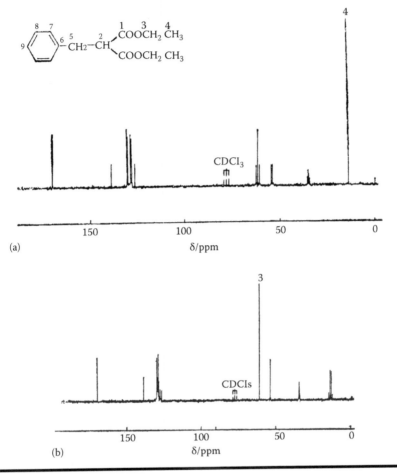

Figure 5.17 **Selective coupled spectra of compound A. (a) On the #4 carbon atom; (b) on the #3 carbon atom. (From Ning, Y.C.** *Structural Identification of Organic Compounds and Organic Spectroscopy,* **2nd ed. Science Press, 2000.)**

(such as low natural abundance of the ^{13}C nucleus, and thus poor resolution; non-uniform excitation of ^{13}C nuclei with different resonance frequencies; and several decoupling techniques are applied to highlight the characteristic structure), a quantitative relationship is not observed between the integrated peak area and number of carbon nuclei. Hence, it is not possible to quantify the number of carbon nuclei in ^{13}C NMR spectroscopy, similar to that carried out in ^{1}H NMR spectroscopy. Instead, chemical shifts are mainly identified.

For reference, Figure 5.18 shows the ranges of chemical shifts for some important carbon nuclei in organic functional groups. The chemical shifts approximately rank from right to left (low to high) as follows: saturated hydrocarbons, unsaturated hydrocarbons, aromatics, carboxylic acids, and ketones.

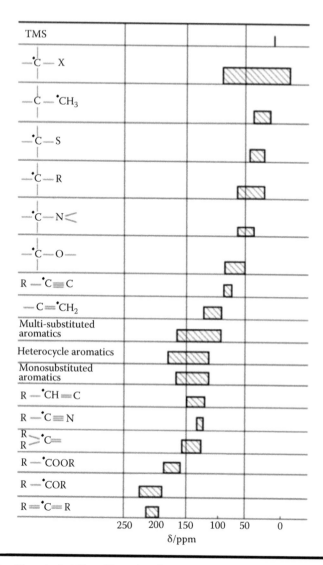

Figure 5.18   Chemical shifts of functional groups in ¹³C NMR spectroscopy (for *C).

5.4.2.4 *Empirical Expressions of ¹³C Chemical Shifts [4,5]*

■ Saturated Hydrocarbon
Grant has proposed the following empirical expression:

$$\delta_{C(k)} = -2.1 + \sum A_i \tag{5.18}$$

Here, k is the target carbon nucleus. A_i is the calibration coefficient of the ith group from k, as given in Table 5.11.

Table 5.11 Empirical Coefficients in Saturated Hydrocarbons Proposed by Grant

^{13}C Nucleus	A	^{13}C nucleus	A
α	9.1	2°(3°)	−2.5
β	9.4	2°(4°)	−7.2
γ	−2.5	3°(2°)	−3.7
δ	0.3	3°(3°)	−9.5
ε	0.1	4°(1°)	−1.5
1°(3°)	−1.1	4°(2°)	−8.4
1°(4°)	−3.4		

Note: Notably, 1°(3°) indicates that the target primary carbon atom is connected to a tertiary carbon atom, while 2°(4°) indicates that the target secondary carbon atom is connected to a quaternary carbon atom.

Lindeman and Adams have proposed another empirical expression. In $\overset{k}{\underset{n}{>}}CH_n - \overset{\alpha}{C}H_m - \overset{\beta}{C} - \overset{\gamma}{C} - \overset{\delta}{C}$, the chemical shift of the k carbon nucleus is expressed as follows:

$$\delta_c(k) = A_n + \sum_{m=0}^{2} N_m^\alpha a_{nm} + N^\gamma \gamma_n + N^\delta \delta_n \tag{5.19}$$

Here, N^α, N^γ, and N^δ represent the number of carbon atoms on α, γ, and δ positions, respectively. n is the number of hydrogen atoms on the target k carbon atom, and m is the number of hydrogen atoms on the α carbon atom, and the other variables are constants. Table 5.12 summarizes the values of these variables [6].

■ Monosubstituted Alkene

The chemical shifts of a monosubstituted alkene $R-\overset{1}{C}H = \overset{2}{C}H_2$ are as follows:

$$\delta_c(k) = 123.3 + \sum_i Z_{ki}(R_i) \tag{5.20}$$

Here, $Z_{ki}(R_i)$ represents the effect of the R_i substituent on the chemical shift of k carbon atoms ($k = 1$ and 2), as shown in Table 5.13.

Table 5.12   Empirical Parameters in the Expression by Lindeman and Adams [6]

n	A_n	m	a_{nm}	γ_n	δ_n
3	6.80	2	9.56	−2.99	0.49
		1	17.83		
		0	25.43		
2	15.34	2	9.75	−2.69	0.25
		1	16.70		
		0	21.43		
1	23.46	2	6.60	−2.07	~0
		1	11.14		
		0	14.70		
0	27.77	2	2.26	0.86	~0
		1	3.96		
		0	7.35		

Source: Lindeman, L.P., and J.Q. Adams, *Analytical Chemistry*, 1971, 43(10), 1245–1252.

Table 5.13   Effects of Substituents in Monosubstituted Alkenes

Substituent	Z_1	Z_2	Substituent	Z_1	Z_2
-H	0.0	0.0	-Ph	12.5	−11.0
$-CH_3$	10.6	−7.9	-Cl	2.6	−6.1
$-CH_2CH_3$	15.5	−9.7	-F	24.9	−34.3
$-CH(CH_3)_2$	20.4	−11.5	$-OCH_3$	29.4	−38.9
$-C(CH_3)_3$	25.3	−13.3	-CHO	13.1	12.7
$-CH_2OH$	14.2	−8.4	$-COCH_3$	15.0	5.8
$-CH=CH_2$	13.6	−7.0	-COOH	4.2	8.9

■ Substituted benzene

The chemical shifts of substituted benzene $_4$⟨◯⟩$_1$—R are expressed as follows:

$$\delta_c(k) = 128.5 + \sum_i Z_{ki}(R_i) \tag{5.21}$$

Here, $Z_{ki}(R_i)$ is the effect of the R_i substituent on the chemical shift of k carbon atoms ($k = 1–4$), as shown in Table 5.14.

Table 5.14   Effect of Substituents in Substituted Benzene

Substituent	Z_1	Z_2	Z_3	Z_4
-H	0.0	0.0	0.0	0.0
-CH$_3$	9.3	0.6	0.0	-3.1
-C(CH$_3$)$_3$	22.1	-3.4	-0.4	-3.1
-CH$_2$OH	13.0	-1.4	0.0	-1.2
-CH=CH$_2$	7.6	-1.8	-1.8	-3.5
-F	35.1	-14.3	0.9	-4.4
-Cl	6.4	0.2	1.0	-2.0
-Br	-5.4	3.3	2.2	-1.0
-OH	26.9	-12.7	1.4	-7.3
-OCH$_3$	30.2	-14.7	0.9	-8.1
-OCOCH$_3$	23.0	-6.4	1.3	-2.3
-NH$_2$	19.2	-12.4	1.3	-9.5
-NHCH$_3$	21.7	-16.2	0.7	-11.8
-N(CH$_3$)$_2$	22.4	-15.7	0.8	-11.8
-NCO	5.7	-3.6	1.2	2.8
-CHO	9.0	1.2	1.2	6.0
-COCH$_3$	9.3	0.2	0.2	4.2
-COOH	2.4	1.6	-0.1	4.8
-COOCH$_3$	2.1	1.2	0.0	4.4
-CONH$_2$	5.4	-0.3	-0.9	5.0
-CN	-16.0	3.5	0.7	4.3

5.4.2.5 High-Resolution Solid-State NMR Spectroscopy

For polymer materials in the solid state, an abundant amount of structural information is closely related to service performance. Nevertheless, as mentioned in Section 5.1.4, the spin–spin relaxation time T_2 is very short in solids; hence, the resonance signal is a wide peak with poor resolution. In addition, various anisotropy phenomena are observed in the solid state, which contribute to peak broadening. For example, protons in water exhibit a peak width of approximately 0.1 Hz, but protons in ice exhibit a peak width of up to 10^5 Hz. Thus, the direct NMR measurement of solids is not suitable to provide information. On the other hand, if a polymer material is dissolved in a solvent, an abundant amount of structural information with respect to the aggregation state (such as crystallization, orientation, and phase separation) cannot be obtained. Hence, solid-state NMR spectra should be recorded. For improving the resolution of spectra, particularly of ^{13}C NMR spectra, some techniques have been developed. By the combination of these below-mentioned techniques, significant improvement in resolution is observed. The following are the typical techniques:

■ Magic Angle Spinning (MAS)

Interaction between nuclei is related to $3\cos^2\beta - 1$, where β is the angle between the magnetic field B_0 and the rotation axis of solid samples. When a sample is rotating at $\beta = 54.74°$ with high speed (typically 30 kHz), $3\cos^2\beta -1 = 0$; hence, dipole interaction between nuclei as well as the anisotropy of chemical shifts disappear, and high-resolution solid-state NMR spectra are recorded. However, because of the low natural abundance of ^{13}C nuclei, only MAS is not sufficient to obtain satisfactory spectra.

■ Cross Polarization (CP)

For overcoming the disadvantage of the low abundance of ^{13}C nuclei, a nucleus with high abundance (e.g., ^{1}H) is polarized first, and its energy is then transferred to ^{13}C nuclei so as to increase their signal intensity.

■ High-power Proton Decoupling (HPD)

A high-intensity, high-power magnetic field is applied to decouple the dipole interactions between ^{13}C and ^{1}H nuclei in solids and improve spectral resolution.

By the combination of these techniques, a spectrum with a significantly improved resolution is obtained (Figure 5.19) [7].

High-resolution solid-state NMR spectroscopy is suitable for insoluble polymers (e.g., crosslinked polymers) and aggregated structures, such as conformation, steric regularity, crystallization, orientation, and phase separation. In addition, relaxation times are helpful for monitoring molecular movement, revealing heterogeneity in polymer materials, and investigating interfacial interactions in composites.

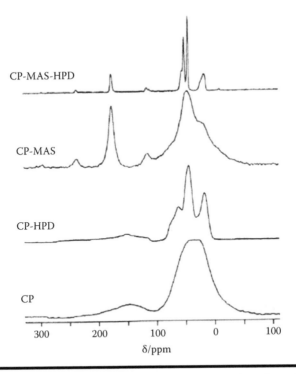

Figure 5.19   Solid-state ^{13}C NMR spectrum of polymethyl methacrylate (PMMA) at 75 MHz. (Harris, R.K., NMR studies of solid polymers, In *Polymer Spectroscopy*, edited by A.H. Fawcett. 1996. Copyright Wiley-VCH Verlag GmbH & Co. KGaA. Reproduced with permission.)

5.5 Applications in Polymer Materials

NMR is a powerful tool for elucidating the structures of polymers, especially for qualitative identification and quantitative analysis, and for monitoring steric regularity and molecular movement.

5.5.1 Identification of Polymers

Structural elucidation by NMR has been discussed before. Apart from typical qualitative analysis based on chemical shifts, peak integration, and coupling, NMR is quite efficient for distinguishing polymers with similar structures, including isomers. For example, polyethyl acrylate:

and polyvinyl propionate:

$$
\begin{array}{c}
\text{---CH}_2\text{---CH---}_n \\
| \\
\text{O---C---}\overset{a}{\text{CH}}_2\overset{b}{\text{CH}}_3 \\
\|\hspace{0.5em} \\
\text{O}
\end{array}
$$

have the same chemical formula of the repeating unit ($C_5H_8O_2$) and similar IR spectra, and their chemical structures can be easily distinguished by ^1H NMR spectroscopy. For example, H_a is coupled with the adjacent CH_3 group and is split into a quartet. H_b is coupled with the adjacent CH_2 group and is split into a triplet. However, they exhibit different chemical shifts. In polyethyl acrylate, H_a exhibits a large chemical shift ($\delta_{H_a} = 4.12$ ppm), caused by the neighboring oxygen atom, while in polyvinyl propionate, H_a exhibits a small chemical shift ($\delta_{H_a} = 2.25$ ppm), corresponding to the neighboring carbonyl group.

5.5.2 *Component Analysis of Copolymers*

In ^1H NMR spectroscopy, the integrated peak area is proportional to the number of protons. Hence, quantitative analysis is possible without the use of standards. For example, in the ^1H NMR spectrum of a butadiene–styrene copolymer, the peak observed at $\delta = 5$ ppm corresponds to the proton of the C=C double bond in the butadiene unit, while the peak observed at $\delta = 7$ ppm corresponds to the proton of the benzene ring.

If A represents the integrated peak area at $\delta = 7$ ppm, and R represents the sum of the areas of other peaks, where a represents the peak area corresponding to a single proton, A/a aromatic protons and R/a aliphatic protons are present in the copolymer. One styrene unit contains five aromatic protons; hence, $A/5a$ styrene units are present. In addition, one styrene unit contains three aliphatic protons; hence, the number of protons in a butadiene unit is expressed as follows:

$$
R/a - \frac{3A}{5a} = \frac{1}{a}\left(R - \frac{3}{5}A\right) \tag{5.22}
$$

The number of butadiene units is expressed as follows:

$$
\frac{1}{6a}\left(R - \frac{3}{5}A\right) = \frac{1}{30a}(5R - 3A) \tag{5.23}
$$

Hence, the molar ratio of styrene to butadiene in the copolymer is expressed as follows:

$$S/B = (A/5a)/\left[\frac{1}{30a}(5R - 3A)\right] = 6A/(5R - 3A) \qquad (5.24)$$

In some cases, a polymer chain contains a special end group, which is different from the repeating units, and the number-average molecular weight can be determined if the protons in the end group and in the repeating units exhibit distinct chemical shifts.

5.5.3 Steric Regularity of Polymers

The steric regularity of polymers significantly affects crystallization and mechanical properties. In Chapter 2, PGC can be employed to differentiate polymers with different steric regularities. NMR is also effective for characterizing the steric regularity of polymers. In various steric isomers, the chemical environment of specific protons may be different; hence, their chemical shifts are different.

PMMA is a typical example for investigating steric regularity. Three protons are present in three spin systems: protons in methylene, methyl, and methoxy groups, respectively. Hence, three singlets are observed in the ^{1}H NMR spectrum of PMMA. However, because of different steric regularities, various peaks are split (Figure 5.20). Let us first examine the CH_2 group. In syndiotactic PMMA, two protons in the CH_2 group are equivalent; hence, a singlet is observed. In isotactic PMMA, one proton is near the CH_3 group, while the other is near the $COOCH_3$ group. The two nonequivalent protons afford a quartet via AB-type coupling. In atactic PMMA, two nonequivalent protons are always present. Their heterocoupling results in various splitting patterns, which in turn leads to several small peaks. Protons in the CH_3 group exhibit a similar coupling splitting. In isotactic and syndiotactic PMMA, three equivalent protons are present in the CH_3 group, corresponding to three singlets. However, their chemical shifts are different, corresponding to different chemical environments in isotactic PMMA and syndiotactic PMMA. In atactic PMMA, three nonequivalent protons are present, resulting in a multiplet. The relative ratio of the three steric isomers can also be obtained via the determination of the relative intensities of *mmmm*, *rrrr*, and heterogeneous *mr* groups.

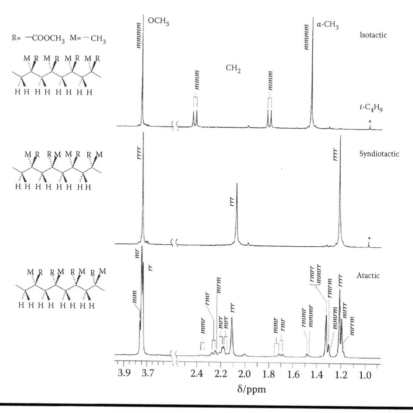

Figure 5.20 ¹H NMR spectrum of PMMA in a nitrobenzene solution (500 MHz, 110°C). The *t*-C₄H₉ peak corresponds to the tert-butyl end group. (With kind permission from Springer Science+Business Media: Hatada, K., and T. Kitayama, *NMR Spectroscopy of Polymers*, 2004.)

5.5.4 Molecular Movement of Polymers

From the principle of NMR, peak width is related to relaxation time. The shorter the relaxation time, the broader the peak. For gases and liquids, T_2 is approximately 1 s. The T_2 of solids is significantly shorter, only 10^{-5} to 10^{-3} s, resulting in very broad peaks. With increasing temperature, molecules move rapidly, and their relaxation times are prolonged, resulting in narrow peaks. For example, during the ¹H NMR measurement of polyisobutylene with increasing temperatures, a step-like change with respect to the peak width is observed [8]. At –90°C, –30°C, and 30–40°C, three abrupt narrowing phenomena are observed, corresponding to the rotation of the CH₃ group at –90°C, segment movement at –30°C, and larger segment movement at 30–40°C, respectively.

References

1. Gunther, H. *NMR Spectroscopy Basic Principles, Concepts and Applications in Chemistry*, 3rd ed. Wiley-VCH: 2013.
2. K Hatada, T Kitayama, *NMR Spectroscopy of Polymers*, Springer-Verlag Berlin Heidelberg, 2004.
3. Ning, Y.C. *Structural Identification of Organic Compounds and Organic Spectroscopy*, 2nd ed. Science Press: 2000.
4. RM Silverstein, FX Webster, *Spectrometric Identification of Organic Compounds* (Sixth Edition), John Wiley & Sons, 1998.
5. R Yang, X Zhou, CQ Luo, KH Wang, *Advanced Intrumental Analysis of Polymers* (Third Edition), Tsinghua University Press, Beijing, 2010.
6. LP Lindeman, JQ Adams, Carbon-13 nuclear magnetic resonance spectrometry—Chemical shifts for paraffins through c9, *Analytical Chemistry*, 1971, 43(10), 1245–1252.
7. RK Harris, NMR studies of solid polymers, in *Polymer Spectroscopy* edited by A.H. Fawcett, John Wiley & Sons, 1996.
8. Spevacek, J., and B. Schneider. "High resolution ^{1}H-NMR relaxation study of poly-isobutylene in solution." *Journal of Polymer Science Part A-2 Polymer Physics* 1976, 14(10), 1789–1800.

Exercises

1. What information can be obtained from an NMR spectrum?
2. Why can NMR spectroscopy be utilized for quantitative analysis? Is a known standard required for quantitative analysis? Why?
3. Can quantitative analysis be carried out by ^{13}C-NMR spectroscopy? Why?
4. If a ^{13}C nucleus is placed in a magnetic field of 2.4 T at 27°C, please calculate the ratio of nuclei in the ground state to the nuclei in the excited state.
5. When NMR spectroscopy is employed for ^{1}H measurement, an RF frequency of 350 MHz is utilized. If the same instrument is employed for ^{13}C, ^{19}F, and ^{31}P measurements, what are the corresponding RF frequencies?
6. A sample exhibits a resonance peak 315 Hz away from the TMS standard at an RF frequency of 60 MHz. What is its chemical shift? If the RF frequency is changed to 100 MHz, what is its chemical shift? Please calculate its distance (in Hz) from TMS.
7. By considering only the first-order rule, please calculate the relative intensities of a, b, c, and d protons. How are the peaks split? (Only ^2J and ^3J are necessary.)

1 CH_3 — OH
    a   b

2 $(CH_3)_3C$ — CH_2Br
    a   b

3 CH_3 — CHC1 — CH_2 — O — CH_3
    a   b   c   d

4 $(CH_3)_2CH$ — O — CH_2 — CH_3
    a   b   c   d

8. The following figure shows the ¹H NMR spectrum of a mixture of benzene, toluene, and dichloromethane. Please calculate the molar ratio of the three components.

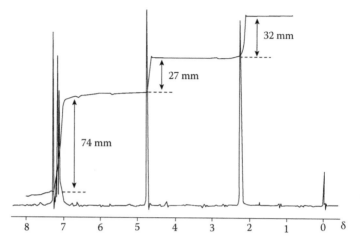

9. In the ¹H NMR spectrum of $CHCl_2–CH_2Cl$, two peaks are observed at δ = 3.95 ppm and δ = 5.77 ppm, respectively, with a coupling constant of 6 Hz. Please draw a ¹H NMR spectrum.

10. Please draw a ¹H NMR spectrum of ethanol, with chemical shifts, relative intensities, and splitting information from J coupling splitting (only ^2J and ^3J are required).

11. Please calculate the chemical shifts of all protons in $C_6H_5CH_2Br$.

12. Please calculate the chemical shifts of protons a, b, c, and d in the following molecules.

(1) ; (2) ; (3)

13. Please calculate the chemical shifts of #1 to #6 carbon nuclei for the following molecule.

14. Only two peaks are observed in the ¹H NMR spectrum of polyisobutylene at δ = 1.46 ppm and δ = 1.08 ppm, respectively. Does it contain any head–head structure? If so, why?

Chapter 6

Thermal Analysis

6.1 Introduction

Thermal analysis involves a group of techniques conducted under controlled temperature programs. Table 6.1 summarizes the different physical properties of polymer materials, such as temperature difference, heat flow, mass, size or volume, deformation, conductivity, and volatile components, which are determined during thermal processes. The controlled temperature program can be linear heating, linear cooling, isothermal processes, thermal cycles, nonlinear heating, or nonlinear cooling as a function of time.

As described in Chapter 2, pyrolysis can also be employed to investigate the changes in polymer materials at high temperature; however, chemical changes, mainly the chain scission of polymer backbones, occur. In thermal analysis, physical changes are mainly examined. Thermal analysis is widely used for both scientific and industrial applications, including polymer materials, foods, pharmaceuticals, electric materials, metals, and ceramics. In this chapter, DSC and TGA are discussed because they are routinely employed for polymer analysis.

6.2 Differential Scanning Calorimetry (DSC)

6.2.1 Principle

Figure 6.1 shows the schematic of a differential scanning calorimetry (DSC) system. Two crucibles, one with a sample and one with a reference, are placed in the sample compartment and are subjected to the same temperature program. Because of different thermal capacities, their responses to the ambient temperature, and thus, their temperatures, are different. A heater in the DSC system heats the

Table 6.1   Thermal Analysis Techniques

Physical Property	Thermal Analysis Techniques
Temperature difference	Differential thermal analysis (DTA)
Heat flow	Differential scanning calorimetry (DSC)
Mass	Thermogravimetric analysis (TGA)
Volatility	Evolved gas analysis (EGA)
Size or volume	Dilatometry
Deformation	Thermomechanical analysis (TMA) Dynamic mechanical analysis (DMA)
Thermal conductivity	Thermal conductivity analysis (TCA)
Electric current	Thermally stimulated current (TSC)
Luminescence	Thermoluminescence (TL)

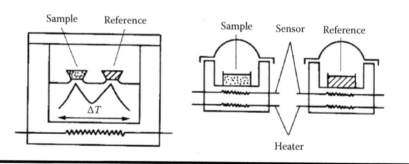

Figure 6.1   Schematic of a DSC system (left: heat-flux-type DSC; right: power-compensation-type DSC).

crucible which temperature is less than that of the other crucible to ensure that both crucibles are at the same temperature. A DSC curve is obtained by plotting the heat flux (dH/dt or dQ/dt), measured as the vertical axis, versus temperature (or time) during the experiment as the horizontal axis (Figure 6.2).

Figure 6.2 shows the DSC curve for a typical polymer, polyethylene terephthalate (PET). The endothermic or exothermic peak, or the baseline jump in the curve, represents the transition temperature of the polymer. The baseline step at 77°C represents the change in the thermal capacity (C_p), which corresponds to the glass transition temperature (T_g) of PET. The exothermic peak observed at 136°C corresponds to the cold crystallization temperature (T_c). The endothermic peak observed at 261°C corresponds to the melting temperature (T_m) of PET crystals. The shaded

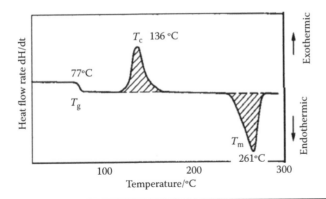

Figure 6.2 **DSC curve of polyethylene terephthalate (PET).**

area represents the heat absorbed or released (phase change enthalpy). From the above discussion, DSC provides considerable information; hence, it is widely employed to investigate polymer materials.

6.2.2 *Instrument and Sample*

1. Instrument

From the principle of DSC, the DSC instrument comprises a furnace, temperature controller, power compensation unit, and data collector. Two crucibles are present in the furnace: one for the sample and the other for the reference. The temperature controller is responsible for controlling the furnace temperature. The power compensation unit is responsible for heating one crucible, the temperature of which is less than that of the other crucible, to ensure that both crucibles are at the same temperature. Besides, a purge gas system is responsible for supplying the necessary atmosphere.

2. Sample

Typically, solid samples are used in DSC. As polymer materials exhibit poor thermal conductivity, sample powders or films are preferred to ensure homogeneous heating. Depending on the sample nature, the sample mass is typically 1–10 mg. If a sample exhibits good thermal conductivity, a relatively large sample amount is allowed. Figure 6.3 shows the effect of sample mass on the melting temperature of polyethylene (PE) [1]. A large sample amount affords a high melting temperature. For composites or polymer blends, a large sample amount is needed for overcoming the heterogeneity and ensuring that the result reflects the actual behavior.

Although DSC typically deals with solids, it can also deal with liquids. In this case, the crucible needs to be annealed.

A DSC sample is added into a crucible, which should be inert in the measurement range to prevent possible reactions with samples or its own thermal

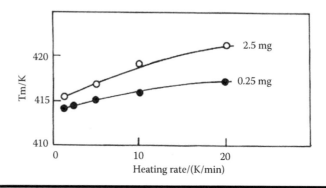

Figure 6.3 **Effect of heating rate and sample amount on the melting tempera-
ture of PE. (Hatakeyama, T., and F.X. Quinn. Thermal *Analysis–Fundamentals and
Applications to Polymer Science.* 1994. Copyright Wiley-VCH Verlag GmbH & Co.
KGaA. Reproduced with permission.)**

transition. Typically, aluminum crucibles are used for measurements conducted at temperatures less than 500°C. On the other hand, alumina or platinum crucibles are used for measurements conducted at higher temperatures.

6.2.3 DSC Techniques

DSC determines temperature and heat. Hence, it is mandatory to routinely calibrate
the temperature and instrument parameters to ensure the precision of results. In the
target temperature range, a straight line (baseline) is obtained under programmed
temperatures for measurements conducted with no samples in the sample and reference crucibles. Standards (Table 6.2) are used for temperature and heat calibration.

Table 6.2 **Standards Used for the Calibration of Temperature and Heat in DSC**

Standard	Melting Temperature/°C	Fusion Enthalpy ΔH_f(J/g)
Azobenzene	34.6	90.43
Stearic acid	69	198.87
Phenanthrene	99.3	104.67
Indium (In)	156.4	28.59
Pentaerythritol	187.8	322.80
Tin (Sn)	231.9	60.62
Lead (Pb)	327.4	23.22
Zinc (Zn)	419.5	111.4

In a majority of commercial DSC instruments, high-purity In (99.999%) is used as the standard for the calibration of temperature and enthalpy.

For polymer materials, DSC analysis is typically carried out at temperatures less than 350°C. At a higher temperature, polymer materials may decompose and contaminate the sample compartment and furnace.

1. Definition of the Thermal Transition Temperature [1].

As shown in the DSC curves in Figure 6.4, the glass transition of polymer materials is indicated as a jump in the baseline in a specific temperature range, where subscripts i and m refer to the initial and middle points, respectively. The glass transition temperature is considered as T_{ig}, which represents the crossover point of the tangent to the upward curve and the baseline. In addition, the glass transition temperature can be considered as T_{mg}, which represents the crossover point of the tangent to the upward curve and the middle line.

The melting or crystallization of crystals affords an endothermic or exothermic peak. Figure 6.5 shows a melting curve. T'_{im}, T_{im}, and T_{pm} are considered the melting temperatures. For polymer materials, the peak temperature T_{pm} is predominantly used. The crystallization temperature can be selected in a similar manner.

2. Scanning Rate

As polymer materials exhibit poor thermal conductivity, the scanning rate significantly affects the results. A rapid scanning rate leads to high sensitivity and sharp peaks. However, the determined transition temperature may be greater than the actual value, as shown in Figure 6.3, because the sample is not representative of the change in ambient temperature. In addition, a rapid scanning rate may cause the overlap of the two peaks placed near each other.

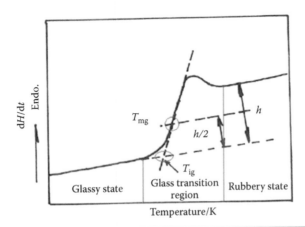

Figure 6.4   A typical DSC curve in the glass transition region.

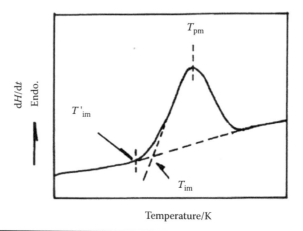

Figure 6.5 Typical DSC curve in the melting region.

In contrast, at a slow scanning rate, the actual sample temperature is equal to the ambient temperature, and the determined transition temperature reflects the actual value. However, measurement is time-consuming. Besides, polymer chains exhibit slow thermal transitions and relaxation at a low scanning rate; hence, clear transition peaks are not observed in the DSC curve. Figure 6.6 shows DSC curves of In at various scanning rates to explain the aforementioned effect. An appropriate scanning rate according to the thermal conductivity of samples must be selected. For polymer materials, a scanning rate of 5–10°C/min is typically employed. For a relatively large sample amount, the effect of the scanning rate is more significant (Figure 6.3).

Although the scanning rate affects the transition temperature, as well as the height and shape of the peaks, it does not affect the endothermic or exothermic peak area, and thus the quantitative analysis of heat flux.

3. Purge Gas

Typically, inert gases, such as N_2 and He, are used as the purge gas. The inert purge gas prevents the oxidation of samples and the corrosion of volatile components in samples to the detector. A high gas flow rate affects the surface temperature of samples, resulting in an unstable baseline. Hence, a purge gas flow rate of 20–40 mL/min is recommended, which should be stably maintained during measurement. When the purge gas is changed, the instrument must be recalibrated. Sometimes, air or O_2 is also used to investigate the oxidation of polymer materials.

4. Thermal History

Thermal history significantly affects the mechanical properties of polymer materials. Particularly, the processing temperature, thermal treatment time, cooling rate, and storage condition of samples results in various crystallization behaviors, which in turn significantly change the thermal transition

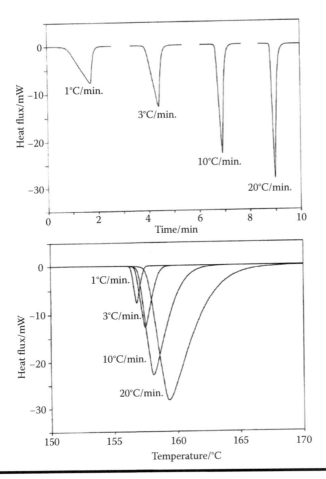

Figure 6.6   Effect of the scanning rate on the DSC curve of indium.

and relaxation of samples in DSC analysis. Hence, considerable attention needs to be focused on analysis and data interpretation. Figure 6.7 shows two DSC curves of a linear low-density polyethylene (LLDPE) sample subjected to two cooling programs. With the cooling of the sample from the melt following the temperature program shown in Figure 6.7a, Curve 1 is obtained for the sample analyzed at a scanning rate of 10°C/min. When the sample is naturally cooled from the melt, Curve 2 is obtained at the same scanning rate. Figure 6.7 clearly shows the effect of thermal history. During step-by-step cooling, it is possible for segments with different lengths to crystallize individually. In the following DSC analysis, these segment crystals melt at elevated temperatures and exhibit a multipeak DSC curve [2].

Another typical example is PET. PET slowly crystallizes, and its crystallization temperature is low; hence, when the cooling procedure is changed,

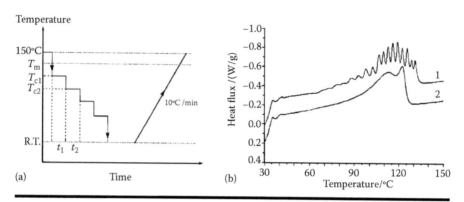

Figure 6.7 **DSC curves of LLDPE subjected to two cooling procedures: Curve 1: cooling by a multistep cooling program in (a) and Curve 2: natural cooling (b).**

the crystallization state changes accordingly, resulting in different behaviors during heating. As shown in Figure 6.8, a weak glass transition and a melting peak are observed in Curve I (original sample). With the cooling of the original sample at a scanning rate of 10°C/min, Curve II is obtained, where a crystallization peak is observed. Then, the sample is heated again at a scanning rate of 10°C/min. A melting peak is also observed in Curve III. However, its peak shape is slightly different from that in Curve I, indicating that its crystallization morphology is slightly different from that of the original sample. Next, the sample is quenched and then heated again, and Curve IV is obtained. This time, a clear glass transition stage, cold crystallization peak, pre-melt

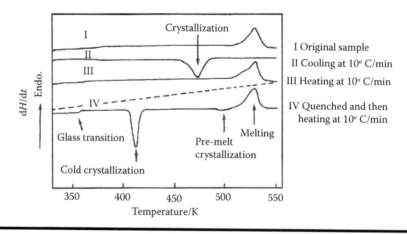

Figure 6.8 **DSC curves of PET after different thermal histories. (Hatakeyama, T., and F.X. Quinn. Thermal *Analysis–Fundamentals and Applications to Polymer Science*. 1994. Copyright Wiley-VCH Verlag GmbH & Co. KGaA. Reproduced with permission.)**

crystallization peak, and melting peak are observed. The quenched sample is roughly amorphous; hence, a clear glass transition is observed. During heating, the sample further crystallizes to afford a cold crystallization peak. With an increase in temperature, a second crystallization occurs, affording a premelt crystallization peak. Finally, when the temperature reaches the melting temperature, a melting peak is observed.

As the thermal history of samples significantly affects DSC curves, this information is mandatory before conducting DSC analysis. If it is not possible, a sample should be heated at a certain scanning rate to a temperature that is 30–50°C greater than the melting point and maintained at that temperature for 5–10 min, followed by cooling the sample at a certain scanning rate to the preset temperature. In this manner, a sample with a known thermal history is obtained.

From the above discussion, several parameters must be considered during DSC analysis because DSC curves and thermal transition temperatures may be affected. Hence, in a DSC analysis report, experimental parameters, such as sample mass, temperature range, scanning rate, and purge gas with its flow rate, must be noted.

6.2.4 Applications in Polymer Materials

1. Glass Transition

The glass transition of polymers in a DSC curve is indicated by a step-like baseline shift. Under a certain pressure, the heat flux rate of a certain sample amount is proportional to the specific heat and scanning rate.

$$dH/dt = (dH/dT)_p (dT/dt) = mC_p dT/dt. \qquad (6.1)$$

At a certain scanning rate, the baseline level reflects the specific heat. During the glass transition of polymers, the specific heat changes. The change of specific heat corresponds to the step height of the baseline shift. In a polymer, the magnitude of the change in the specific heat depends on the amorphous phase content. In a completely crystallized polymer, nearly no baseline shift is detected. Conversely, in a completely amorphous polymer, the highest baseline shift is detected.

The glass transition temperature is the deciding factor for evaluating if a polymer material is a plastic or rubber at room temperature. The glass transition temperature of rubbers is lower than the room temperature; hence, rubbers exist in a typical rubbery state. The glass transition temperature of amorphous thermoplastics ranges from the room temperature to approximately 160°C, while that of heat-resistant thermoplastics is typically greater than 160°C. Hence, it is possible to judge the type of material and obtain

Table 6.3   Glass Transition Temperatures of Some Typical Polymers

Rubber	$T_g/°C$	Plastics	$T_g/°C$	Engineering Plastics	$T_g/°C$
SBR	−61.1	PC	148	PTFE	330
NR	−62.7	PVB	72.6	PVDF	177.7
EPDM	−51.7	PMMA	124.4	PEEK	342.4
NBR	−36.7				
Silicone rubber	−120.1				

some knowledge about its basic performance via the determination of its glass transition temperature. Table 6.3 summarizes the glass transition temperatures of some typical polymers.

DSC can also be employed to investigate the curing of thermoset resins. Figure 6.9 shows the DSC curves of an epoxy resin during two subsequent heating procedures. During the first heating, a small baseline shift and a strong exothermic peak are detected, representing the melting of low-molecular-weight epoxy resin and curing, respectively. Curing is an exothermic reaction, and the area corresponding to this exothermic peak represents the heat of the curing reaction. During the second heating, only

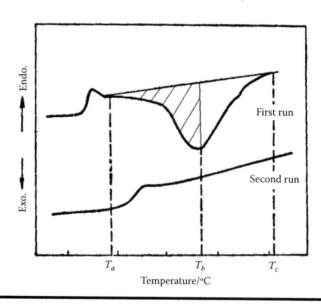

Figure 6.9   DSC curves of epoxy resin during two subsequent heating procedures.

a baseline shift is observed, representing the glass transition of the cured epoxy resin. An exothermic peak is not observed, indicating that curing is completed during the first heating. Hence, the exothermic peak area during the first heating corresponds to a degree of reaction of 100%. Based on the exothermic peak, the curing kinetics can be investigated. For example, in the range of T_a-T_c, the reaction degree at any temperature T_b is equal to the ratio of the integrated area in the range of T_a-T_b (dashed area) to the total peak area.

2. Crystallization or Melting

Several polymers undergo crystallization. Melting temperature is a characteristic parameter and can be used for qualitative identification. Table 6.4 summarizes the melting temperatures and melting enthalpy $\Delta H_{100\%}$ (the extrapolated melting enthalpy for a 100% crystallized polymer from previous studies) of some polymers [3].

In a blend of two polymers with poor compatibility, the two components exhibit their own melting temperatures. On the other hand, if two polymers are compatible, their melting temperatures approach each other or even merge to a single melting peak. Figure 6.10 shows the DSC curves of a series of blends of styrene-p-fluorostyrene copolymer (PFS) and polyphenyl ether (PPE) [4]. At a PFS content of less than 56%, only one melting peak is observed in DSC curves, demonstrating good compatibility between these two components. At PFS contents of 67% and 78%, the melting peaks of these two components in Curves 8 and 9 are completely separated, demonstrating poor compatibility.

Table 6.4 Melting Temperature and Melting Enthalpy $\Delta H_{100\%}$ of Some Polymers

Polymer	Melting Temperature/°C	$\Delta H_{100\%}/(J/g)$
LDPE	110	293.6
HDPE	135	293.6
PP	165	207.1
POM	180	326.2
PA 6	225	230.1
PET	255	140.1

Source: Wagner, M. *Thermal Analysis in Practice* (translated by Liming Lu). Shanghai: Donghua University Press, 2010, p. 73.

Note: Notably, in actual cases, it is not possible to obtain a polymer with 100% crystallinity. $\Delta H_{100\%}$ is obtained by extrapolation.

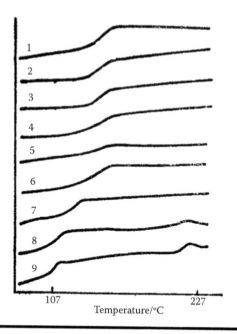

1

2

3

4

5

6

7

8

9

107   227

Temperature/°C

Figure 6.10   DSC curves of a series of blends of styrene-p-fluorostyrene copo-lymer (PFS) and polyphenyl ether (PPE). (From Dong, Y. *Handbook of Polymer Analysis*. Beijing: China Sinopec Press, 2004.) PFS contents in Curves 1–9 are 8%, 16%, 25%, 36%, 46%, 49%, 56%, 67%, and 78%, respectively.

The melting behavior during the heating of polymer, as well as its crys-tallization behavior during cooling, can be investigated by DSC. The area of the melting peak reflects the total heat required to destroy the crystal structure; hence, it represents the fusion heat, which can be used to calculate crystallinity:

$$X_c = \Delta H_f / \Delta H_f^0. \tag{6.2}$$

Here, ΔH_f and ΔH_f^0 represent the fusion heats of the sample and the cor-responding polymer with 100% crystallinity, respectively. The latter can be found in previously published studies or extrapolated from DSC curves using a series of samples with known crystallinity.

DSC is employed to not only measure the melting or crystallization tem-perature and enthalpy but also study the crystallization kinetics of polymers. A typical isothermal crystallization procedure is conducted as follows: first, a sample is heated to a temperature greater than its melting point, which is then allowed to remain at this temperature for several minutes, such that its

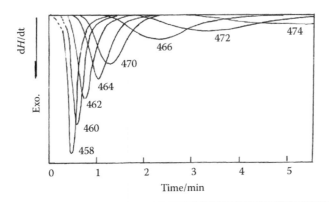

Figure 6.11   Isothermal crystallization curves of carbon-fiber-reinforced nylon 6 at various temperatures (K). (Hatakeyama, T., and F.X. Quinn. Thermal *Analysis–Fundamentals and Applications to Polymer Science*. 1994. Copyright Wiley-VCH Verlag GmbH & Co. KGaA. Reproduced with permission.)

thermal history is normalized; next, it is rapidly cooled to a set crystallization temperature, and the change in the heat flux rate with time is recorded, affording an isothermal crystallization kinetic curve. Figure 6.11 shows the isothermal DSC curves of carbon-fiber-reinforced nylon 6 at various temperatures. At any temperature, the crystallinity at time t is calculated according to Equation 6.2. Here, ΔH_f refers to the fusion heat with time t, i.e., integrated area in the time region of $0-t$.

Typically, the Avrami equation is utilized to describe the crystallization kinetics [5]:

$$1 - X_t = \exp(-Zt^n). \tag{6.3}$$

Here, X_t is the crystallinity at time t; Z is the crystallization rate constant; and n is the Avrami index, which is related to the mechanism of nucleation and growth. Figure 6.12 plots lg $[-\ln (1 - X_t)]$ of carbon-fiber-reinforced nylon 6 versus lg$t$. From the intercept and slope, Z and n can be obtained. From the results, at the start, the crystallization nucleation and growth follow the Avrami equation quite well. However, in the later stage, deviations from the Avrami equation is observed, caused by the contact between large spherulites.

The Avrami index n represents the nucleation and growth mechanism (Table 6.5).

3. Oxidation Induction Time (OIT)

Oxidation induction time (OIT) reflects the thermal stability of polymer materials; hence, it is a crucial parameter in industry. Typically, OIT

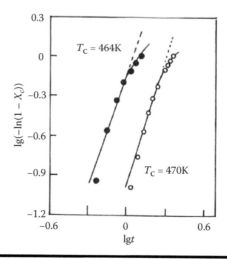

Figure 6.12 lg [−ln(1 − X_t)] versus lgt of carbon-fiber-reinforced nylon 6. (Hatakeyama, T., and F.X. Quinn. Thermal *Analysis–Fundamentals and Applications to Polymer Science*. 1994. Copyright Wiley-VCH Verlag GmbH & Co. KGaA. Reproduced with permission.)

Table 6.5   Avrami Index *n* of Various Nucleation and Growth Processes

Homogeneous Nucleation		Inhomogeneous Nucleation	
Linear Growth	Diffusion-Controlled Growth	Linear Growth	Comment
2	3/2	$1 \leq n \leq 2$	One-dimensional growth
3	2	$2 \leq n \leq 3$	Two-dimensional growth
4	5/2	$3 \leq n \leq 4$	Three-dimensional growth

measurement is carried out by DSC as follows: First, the sample is heated to the set temperature under inert atmosphere (typically N_2). When the baseline levels are off, the purge gas is switched to an oxidative atmosphere (e.g., air or O_2). After some time, the DSC curve departs from the baseline, caused by exothermic oxidation. OIT is defined as the time from the switch to the departure (Figure 6.13). The longer the time, the better the thermal stability of polymer materials.

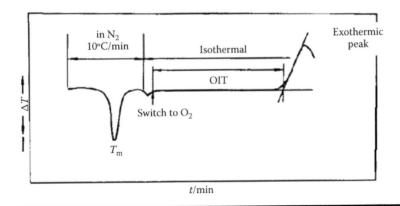

Figure 6.13 OIT measurement of PE.

6.3 Thermogravimetric Analysis

6.3.1 Principle

Thermogravimetric analysis (TGA) measures the change of sample mass with temperature (or time) under programmed temperature (typically heating). Figure 6.14 shows a typical TGA curve, where the horizontal and vertical axes represent temperature T (or time t) and mass loss percentage (or mass), respectively. Typically, a polymer during heating exhibits a step-like mass loss. A TGA curve may have one mass loss step or multiple steps. Mass loss percentages are calculated from the height of each step. The initial mass loss temperature T_i and the final mass loss temperature T_f represent the temperatures at which the mass begins to decrease and

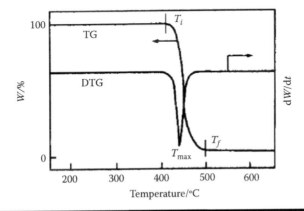

Figure 6.14 A typical TGA curve.

stops to decrease, respectively. In Figure 6.14, the DTG curve is the differential TGA curve. The peak temperature in the DTG curve is T_{max}, corresponding to the maximal weight loss rate.

6.3.2 Instrument and Sample

A TGA instrument comprises a thermal balance, temperature controller, and data collector. A sample is placed in a crucible and heated via a temperature program, and a thermal balance is used to measure the change of mass with temperature.

As polymer samples exhibit poor thermal conductivity, a sample powder (the finer, the better) or a film is appropriate, which is similar to that required in DSC, to ensure that the sample temperature increases with the atmospheric temperature, demonstrating no lag in the sample transition. Typically, a 2–5 mg sample is used. However, for polymer composites or blends, more of a sample (5–10 mg) is used to avoid the effect of heterogeneity.

TGA can attain the maximum temperature of 900–1000°C; hence, crucibles must be resistant to high temperatures, as well as inert to samples, intermediates, final products, and the purge gas. Typically, platinum and alumina crucibles are used.

6.3.3 TGA Techniques

1. Data Processing of TGA [1]

According to the recommendation of ISO, the characteristic weight loss temperature is defined as follows:

(a) One-stage Weight Loss Curve

In a one-stage weight loss curve (Figure 6.15a), the starting point A and ending point B represent the crossover of the tangent of the weight loss and the extension line of the beginning baseline or the crossover of

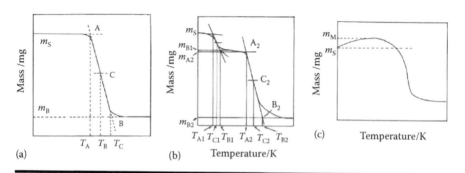

Figure 6.15 Characteristic temperatures in a one-stage weight loss curve (a); a multistage weight loss curve (b); and a mass increase curve (c).

the tangent of the weight loss and the extension line of the final base-line. Point C is the midpoint of A and B. The temperatures at points A, B, and C represent the initial decomposition temperature T_A, middle decomposition temperature T_C, and final decomposition temperature T_B, respectively. The masses at points A and B are m_S and m_B, respectively. The weight loss ratio M_L is expressed as follows:

$$M_L = \frac{m_S - m_B}{m_S} \times 100. \tag{6.4}$$

(b) **Multistage Weight Loss Curve**

Transition points are defined as similar to those defined for the one-stage weight loss curve (Figure 6.15b). The weight loss ratio at each stage is expressed as follows:

$$M_{L1} = \frac{m_S - m_{B1}}{m_S} \times 100. \tag{6.5}$$

$$M_{L2} = \frac{m_{A2} - m_{B2}}{m_S} \times 100. \tag{6.6}$$

(c) **Mass Increase**

Sometimes, mass increase is detected on account of oxidation (Figure 6.15c). The weight increment is expressed as follows:

$$M_G = \frac{m_M - m_S}{m_S} \times 100. \tag{6.7}$$

2. Effect of Experimental Parameters

The heating rate significantly affects the TGA results. As polymer materials exhibit poor thermal conductivity, a moderate heating rate of 5–10°C/min is typically used. A high heating rate results in a high characteristic temperature being detected, since samples cannot exhibit simultaneous responses. In addition, the purge gas affects TGA results. If a sample is investigated under an inert purge gas (typically N_2), thermal decomposition occurs. On the other hand, when a sample is investigated under an oxidative purge gas (O_2 or air), thermal oxidation occurs. The flow rate of the purge gas is approximately 40 mL/min. A high flow rate results in an unstable thermal balance, which aids in transferring heat and conducting the volatile decomposition of the products.

As the results obtained from TGA are closely related to experimental parameters (e.g., sample mass, purge gas, and flow rate, as well as the temperature program), experimental parameters must be reported together with the TGA results.

6.3.4 Applications in Polymer Materials

In principle, all mass changes during a thermal process can possibly be investigated by TGA, including component analysis, thermal property, thermal decomposition and its mechanism, volatile analysis, sublimation and evaporation, oxidation, and reaction kinetics.

1. Thermal Stability Evaluation of Polymers

    The thermal stability of polymers can be evaluated by the comparison of their initial decomposition temperatures or T_{max} in TGA curves determined under the same conditions. Figure 6.16 shows the TGA curves of five polymers—PVC, PMMA, PE, Polytetrafluoroethylene (PTFE), and polyimide (PI). For PVC, the first weight loss is observed from 200°C, and the second weight loss is observed at approximately 400°C. The first and second weight loss peaks correspond to the elimination of side groups and HCl emitted (described in Chapter 2) and the backbone fraction of the main chain, respectively. PMMA, PE, and PTFE exhibit complete decomposition at 400°C, 500°C, and 600°C, respectively. PI is the most stable polymer, which starts to decompose from 600°C, and at 800°C, as much as approximately 50% of the residue remains. Hence, the thermal stability follows the order of PI > PTFE > PE > PMMA > PVC.

    In TGA, the temperatures at which 1% or 5% mass losses, i.e., $T_{1\%}$ or $T_{5\%}$, respectively, are typically utilized to reflect the thermal stability of polymers. Table 6.6 shows the $T_{1\%}$ values of some common polymers in air.

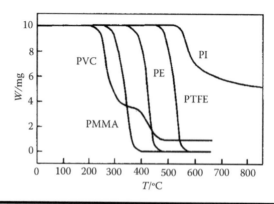

Figure 6.16   TGA curves of five polymers.

Table 6.6 Comparison of Thermal Stabilities of Some Polymers

Polymer	$T_{1\%}$, °C	Polymer	$T_{1\%}$, °C	Polymer	$T_{1\%}$, °C
HDPE	318	PS	330	PVDF	410
LDPE	275	ABS	284	PTFE	502
PP	315	PVA	106	Cotton	215
PB	234	PVC	184	Wool	190
PIP	240				

2. Component Analysis of Polymer Material

If polymers comprise low-molecular-weight additives and inorganic fillers, these components can be easily distinguished from a polymer matrix by TGA. For example, Figure 6.17 shows the TGA curve of a glass-fiber-reinforced resin. At temperatures of less than 200°C, a 2% mass loss is observed, corresponding to the evaporation of a small amount of absorbed water. At 400–600°C, an approximately 80% weight loss is observed, corresponding to the decomposition of the resin matrix, while the final 18% residue corresponds to the glass fiber. Another example is shown in Figure 6.18 for a PTFE composite. At the start, an approximately 0.5% mass loss is observed, corresponding to volatiles in the composite. The next mass loss of 31% corresponds to the decomposition of PTFE. After that, the purge gas is switched from N_2 to air, and then a mass loss of 18% is observed, corresponding to

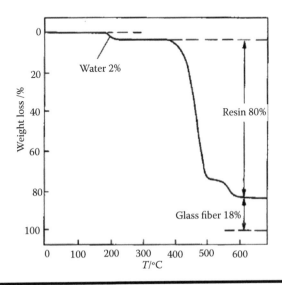

Figure 6.17 TGA curve of glass-fiber-reinforced resin.

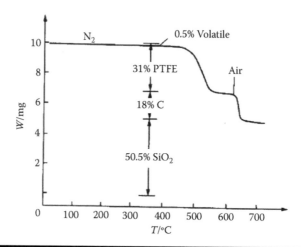

Figure 6.18   TGA curve of PTFE composite.

the oxidation of carbon fillers to CO_2. The final 50.5% residue is SiO_2, which does not decompose irrespective of the surrounding atmosphere being N_2 or air.

In some polymers, the elimination of side groups first occurs to form small molecules, and the scission of the main chain occurs during subsequent heating. Hence, two weight loss stages are detected. For example, the first mass loss in Figure 6.19 corresponds to the elimination of acetic acid from EVA. From the first mass loss, the VA content can be calculated. A two-stage weight loss

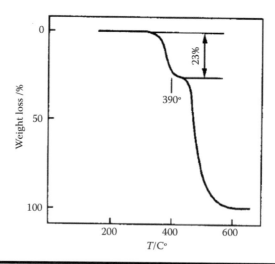

Figure 6.19   TGA curve of EVA.

curve can also be detected in PVC, PVA, and EVOH, where the elimination of side groups can easily occur.

3. Thermal Decomposition Kinetics of Polymer

TGA is a powerful tool for investigating the thermal decomposition kinetics of polymers, which is based on some assumptions: 1) the polymer sample is homogeneous; 2) there is no overlap or parallel thermal decomposition reactions; 3) there is no temperature gradient in the sample; and 4) the diffusion resistance of gases by thermal decomposition can be neglected. Hence, the conversion of thermal decomposition is a function of temperature (or time):

$$\frac{d\alpha}{dt} = k(T)(1-\alpha)^n. \tag{6.8}$$

Here, α is the conversion. $\alpha = \dfrac{m_0 - m}{m_0 - m_\infty}$, m_0, m, and m_∞ represent the masses at the start, at time t, and at the end, respectively. n is the reaction order, which is considered to be constant during thermal decomposition. $k(T)$ is the reaction rate constant, which follows the Arrhenius equation:

$$k(T) = A\exp\left(-\frac{E}{RT}\right). \tag{6.9}$$

Here, A and E are the pre-exponential coefficient and activation energy, respectively.

By substituting Equation 6.9 into Equation 6.8:

$$\frac{d\alpha}{dt} = A\exp\left(-\frac{E}{RT}\right)(1-\alpha)^n. \tag{6.10}$$

At a constant temperature, for a one-order reaction, $n = 1$, then:

$$-\ln(1-\alpha) = kt \tag{6.11}$$

Substituting α and t into Equation 6.11, the thermal decomposition rate constant k can be calculated. Then, the activation energy can be calculated by k at elevated temperatures.

At programmed temperatures, $dT = \beta dt$, and β is the heating rate. Then:

$$\frac{d\alpha}{(1-\alpha)^n} = \frac{A}{\beta}\exp\left(-\frac{E}{RT}\right)dT \tag{6.12}$$

Typically, several methods are utilized to solve Equation 6.12:
(1) Differential Method
  - Freeman–Carroll method [6]:

$$\frac{d\left[\ln(d\alpha/dt)\right]}{d\left[\ln(1-\alpha)\right]} = n - \frac{E}{R}\frac{d(1/T)}{d\left[\ln(1-\alpha)\right]}. \tag{6.13}$$

Hence, by plotting $\dfrac{d\left[\ln(d\alpha/dT)\right]}{d\left[\ln(1-\alpha)\right]}$ versus $\dfrac{d(1/T)}{d\left[\ln(1-\alpha)\right]}$, E and n can be obtained from the slope and intercept, respectively; this method is convenient. Kinetic parameters are obtained only from one TGA curve, but the results are affected by the sample mass and heating rate.
  - Kissinger Method [7]:

$$\ln\left(\frac{\beta}{T_P^2}\right) = \ln\left(\frac{AR}{E}\right) + \ln\left[n(1-\alpha_P)^{n-1}\right] - \frac{E}{RT_P}. \tag{6.14}$$

Here, T_P and α_P are the absolute temperature and conversion at the maximum thermal decomposition rate, respectively. By plotting $\ln\left(\dfrac{\beta}{T_P^2}\right)$ versus $\dfrac{1}{T_P}$, E can be obtained from the slope.

(2) Integral Method
  - Doyle Method [8]:

$$\int_0^\alpha \frac{d\alpha}{(1-\alpha)^n} = \frac{A}{\beta}\int_{T_0}^T e^{-E/RT}\, dT \tag{6.15}$$

$$F(\alpha) = \int_0^\alpha \frac{d\alpha}{(1-\alpha)^n} = \begin{cases} -\ln(1-\alpha) & (n=1) \\ \dfrac{(1-\alpha)^{1-n}}{n-1} & (n \geq 2) \end{cases} \tag{6.16}$$

$$\lg \beta = \lg \frac{AE}{RF(\alpha)} - 2.315 - 0.4567\frac{E}{RT} \tag{6.17}$$

By obtaining TGA curves at different heating rates, at a certain conversion, plotting $\lg \beta$ versus $\dfrac{1}{T}$, activation energy E can be obtained from the slope.

- Coats–Redfern Method [9]:

$$\ln\frac{F(\alpha)}{T^2} = \ln\frac{AR}{\beta E} - \frac{E}{RT}. \tag{6.18}$$

From $\ln\dfrac{F(\alpha)}{T^2} \sim \dfrac{1}{T}$ curve, E and A are calculated.

- Ozawa Method [10]:

$$\ln(1-\alpha)^n \cong \ln\frac{AE}{R} - \ln\beta - 5.33 - 1.05\frac{E}{RT}. \tag{6.19}$$

At different heating rates, at a certain conversion,

$$\ln\beta \cong \text{constant} - 1.05\frac{E}{RT}. \tag{6.20}$$

This method can be widely used to investigate the decomposition kinetics of polymers, but the thermal decomposition function is assumed to be independent of thermal history.

Figure 6.20 shows the thermal decomposition kinetic curve of poly(p-phenylene benzobisthiazole) (PBO) by the Kissinger method. The activation energy is 352.2 kJ/mol. Figure 6.21 shows the decomposition kinetic curves of PBO at various conversions by the Doyle method. The average activation energy is 338.3 kJ/mol [11]. The results by these two methods are in good agreement.

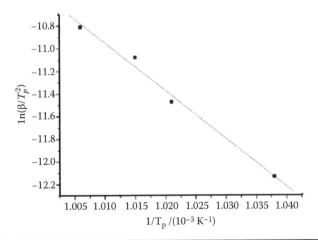

Figure 6.20 Thermal decomposition kinetic curve of PBO by the Kissinger method.

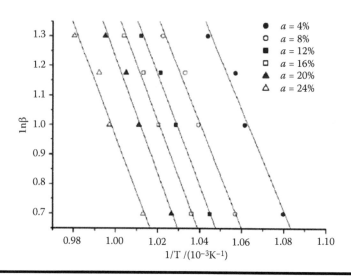

Figure 6.21 **Thermal decomposition kinetic curves of PBO at various conversions by the Doyle method.**

6.4 Coupling Techniques of Thermal Analysis

With the development of thermal analysis techniques, more and more coupling techniques between thermal analysis methods, or with other qualitative and quantitative methods, have been developed and utilized.

Endothermic or exothermic behavior and mass can be simultaneously measured from synchronous TGA–DSC. For example, for the thermal decomposition of a sample, the mass loss is accompanied by the endothermic effect. For the burning of a sample, the mass loss is accompanied by the exothermic effect.

Both TGA-MS and TGA-FTIR determine volatile components by thermal decomposition. In TGA–MS, a fused silica capillary column is inserted between TGA and MS, which must be heated to approximately 200°C to prevent gaseous components from condensation. Only approximately 1% gas from TGA enters into MS for qualitative analysis. The purge gas is Ar or N_2. In TGA–FTIR, 100% gas enters into the gas cell of an FTIR via a warmed capillary column, and it is simultaneously detected. The purge gas is N_2.

References

1. Hatakeyama, T., and F.X. Quinn. Thermal *Analysis–Fundamentals and Applications to Polymer Science*. Chichester, London: John Wiley & Sons, 1994.
2. Teng H. X., Shi Y., and Jin X. G. Novel characterization of the crystalline segment distribution and its effect on the crystallization of branched polyethylene by differential scanning calorimetry. *Journal of Polymer Science Part B: Polymer Physics*, 2002, 40(18), 2107–2118.

3. Wagner, M. *Thermal Analysis in Practice* (translated by Liming Lu). Shanghai: Donghua University Press, 2010, 73.
4. Dong, Y. *Handbook of Polymer Analysis*. Beijing: China Sinopec Press, 2004.
5. Charsley, E.L., and S.B. Warrington. *Thermal Analysis–Techniques and Applications*. Cambridge: The Royal Society of Chemistry 1992.
6. Freeman, E. S, B. Caroll. The application of thermoanalytical techniques to reaction kinetics. The thermogravimetric evaluation of the kinetics of the decomposition of calcium oxalate monohydrate, *Journal of Physical Chemistry*, 1958, 62(4), 394–397.
7. Kissinger, H. E. Reaction kinetics in differential thermal analysis, *Analytical Chemistry*, 1957, 29(11), 1702–1706.
8. Doyle, C. D. Kinetic analysis of thermogravimetric data, *Journal of Applied Polymer Science*, 1961, 5(15), 285–292.
9. Coats, A. W., and J. P. Redfern. Kinetic parameters from thermogravimetric data, *Nature*, 1964, 201(491), 68–69.
10. Ozawa, T. A new method of analyzing thermogravimetric data, *Bulletin of the Chemical Society of Japan*, 1965, 38(11), 1881.
11. Lin, H., Q. Zhuang, J. Cheng, Z. Liu, and Z. Han. Kinetics of thermal degradation of Poly(p-phenylene benzobisoxazole), *Journal of Applied Polymer Science*, 2007, 103(6), 3675–3679.

Exercises

1. When determining the melting temperature and the fusion heat of a crystallized polymer, what factors affect the results? Please explain from the perspective of the effect of polymer structure and DSC instrument parameters.

2. If the DSC curves of the same uncured epoxy resin are obtained for three cycles of continuous heating–cooling, what do the three DSC curves during heating resemble? What information can be obtained from these curves?

3. In the TGA analysis of EVA in N_2, if the mass loss corresponding to the first stage is 10.2%, please calculate the VA content in EVA.

4. Which factors affect TGA results? Please explain.

5. Is it possible to study combined water in crystal, absorbed water, and free water in an inorganic salt by thermal analysis? Why?

6. Is it possible to study the flame retardancy of polymer materials by thermal analysis? Why?

Chapter 7

Microscopic Analysis

7.1 Introduction

Although various analytical instruments can be used to obtain information on the microstructure of polymer materials, visualizing the structure of a sample is advantageous over dealing with a collection of numerical and spectral data. However, the human eye can only resolve objects up to ~0.2 mm. Smaller objects are observed using microscopes. The resolution limit of a typical light microscope is ~0.2 μm, which is 1000× than that of a naked eye. Electron microscopies can resolve length scales smaller than 0.2 nm, which is 1,000,000× the resolution of a naked eye. Owing to the rapid development of microscopy techniques, observations on many different scales have become possible, which significantly contributed to the development of nanomaterials and nanotechnology.

The length scales of various structures in polymer materials are listed in Table 7.1.

7.2 Light Microscopy

The maximal magnification of a typical light microscope is 1000–1500×, corresponding to the resolution limit of ~0.2 μm. Light microscopy is therefore suitable for observations of crystal morphology, crystallization processes, and orientation in polymer materials, blends or block copolymers, multiphase structures of composites, thin films, and fibers, to name a few.

The magnification of a light microscope M is defined as

$$M = M_1 \times M_2 \tag{7.1}$$

Table 7.1 Length Scales of Structures in Polymer Materials

Scale	Structure in Polymer Materials
1–10 nm	Macromolecular coils, crystal nuclei, amorphous and crystal domains
10 nm–1 μm	Isolated domains in multiphase systems (e.g., ABS), emulsions, pigments, filler particles, inorganic nucleating agents, microfibers
1–10 μm	Aggregates of pigments or fillers, spherulites, sectional morphology (especially in fiber reinforced plastics), glass fibers
10 μm–1 mm	Foam structures, textile structures, coating structures

Here, M_1 and M_2 are the magnifications of the objective and the eyepiece.

The resolving power of a light microscope is affected by the diffraction of light. Light rays from an ideal point source pass through a lens and an Airy disk, not a point, is observed in the image plane. The Airy disk consists of a central disk and a series of coaxial rings, as shown in Figure 7.1. The size of the Airy disk is denoted by R_d, and the radius of the first dark ring is:

$$R_d = \frac{0.61\lambda}{n \sin \alpha} M \tag{7.2}$$

Here, λ is the wavelength of the light from point source A; n is the refractive index of the medium on the object side; α is the aperture semi-angle of the lens, i.e., half of the angular aperture of the lens.

A specimen contains many object points, each of which can be regarded as a point source and has its Airy disk in the image plane. If two points are close, their Airy disks may overlap partially. Let us consider two Airy disks with the same size. When the distance between the disks is equal to the radius of the first dark ring, the intensity in the overlapped region is by 20% lower than the central intensity of each Airy disk (Figure 7.1b). Now, R_d is the minimal distance between the centers of two barely resolved disks. Accordingly, the distance between two object points Δr_d is defined as the resolution:

$$\Delta r_d = \frac{R_d}{M} \tag{7.3}$$

Substituting Equation 7.2 into Equation 7.3, we obtain:

$$\Delta r_d = \frac{0.61\lambda}{n \sin \alpha} \tag{7.4}$$

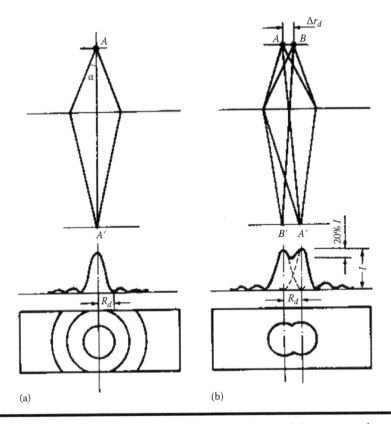

Figure 7.1 **(a) Airy disk of a point source (b) and the resolving power of an optical lens.**

For light microscopes, the maximal aperture semi-angle α is in the 70°–75° range. If the medium on the object side is an oil, with $n \approx 1.5$, then:

$$\Delta r_d \approx \frac{1}{2}\lambda \tag{7.5}$$

That is to say, the resolution limit of a typical light microscope is half of the illumination wavelength. The visible range wavelengths are 390–760 nm; thus, a typical light microscope cannot resolve below 200 nm.

Polarization microscopy and phase contrast microscopy are often used to observe crystal morphology and heterogeneous morphology in polymer materials.

7.2.1 Crossed Polarization Microscopy

In crossed polarization microscopy (CPM), two polarizing sheets (a polarizer and an analyzer) are vertically placed above and under the specimen stage of a light

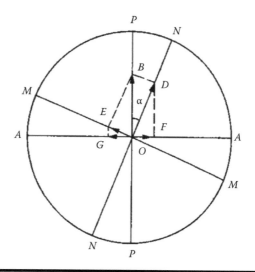

Figure 7.2 Functions of polarizer P–P and analyzer A–A.

microscope, and a specimen is placed between them. CPM is suitable for studying the spherulite structure and the orientation degree of polymer materials.

Figure 7.2 explains interferences in a system of a polarizer P–P and an analyzer A–A. N–N and M–M are two vibration directions in a crystal.

The illuminating light passes through P–P and changes to plane-polarized light along the P–P direction, with vibration amplitude of OB. The specimen splits the beam into two components, i.e., OE along M–M, and OD along N–N. When the beam arrives at A–A, it is split again to OG and OF. So, the transmitted light is

$$\gamma = OF - OG = OD \sin \alpha - OE \cos \alpha \tag{7.6}$$

For an isotropic specimen, the illuminating light cannot pass through vertical polarizing sheets, and so a black field is observed. For an anisotropic crystallized or oriented specimen, the illuminating light is split into two vertical beams with different velocities and refractive indices, and so bright birefringence (cross extinction) is observed.

Figure 7.3 shows the CPM images of poly(l-lactic acid) spherulites. The ordinary image is black and white. If a compensator is used, chromatic images are obtained. For example, the interference color of a first-order red glass slide is magenta. With a first-order red glass slide in the 45° direction, if the first and the third quadrants in the image become blue, while the second and the fourth quadrants become yellow, the specimen is considered to be a positive spherulite (the radial refractive index is higher than the tangential refractive index). Conversely, the specimen is considered to be a negative spherulite. It is clear that poly(L-lactic acid) is a negative spherulite.

Figure 7.3 Poly(l-lactic acid) spherulite after isothermal crystallization at 120°C. Left: without a compensator. Right: with a first-order red glass slide. (Courtesy of Jun Xu, Department of Chemical Engineering, Tsinghua University.)

If a temperature-controlled specimen stage is equipped with a CPM, *in situ* observations of crystallization/melting processes in polymer materials can be performed. The spherulite size can be monitored over time, and the crystallization rate can be calculated. In addition, polymorphism can be observed. For example, Figure 7.4 shows the crystallization process of a PP melt, in which an α-crystal forms first, followed by a β-crystal.

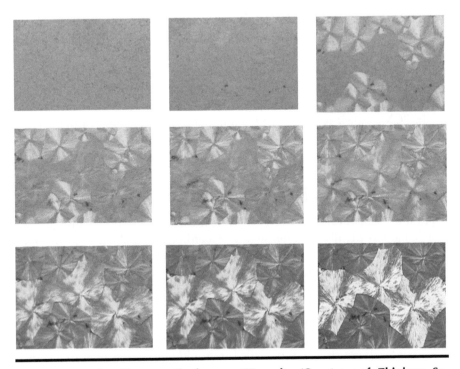

Figure 7.4 Spherulite growth from a PP melt. (Courtesy of Zhiqiang Su, Department of Chemical Engineering, Tsinghua University.)

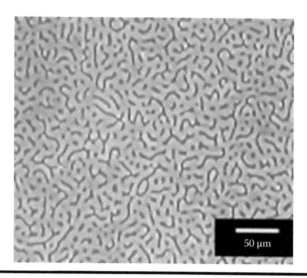

Figure 7.5 **Microphase separation structure of a polymer. (Courtesy of Xuming Xie, Department of Chemical Engineering, Tsinghua University.)**

7.2.2 Phase Contrast Microscopy

Most polymer materials are light-colored or colorless, making it difficult to distinguish them in a multiphase system, i.e., the contrast is poor. A phase contrast microscope (PCM) maps subtle differences between the refractive indices of components onto contrast variations, allowing to distinguish various components and phases. Figure 7.5 shows the microphase separation structure of a polymer.

In PCM, surface roughness and non-uniform thickness also contribute to the contrast, which should be accounted for during sampling.

7.3 Transmission Electron Microscopy

Light microscopes cannot resolve length scales below 200 nm because they use visible light as the illumination source. According to Equation 7.4, the resolution of microscopy can be improved by using shorter wavelengths for illuminations. Electron microscopy (EM) uses electron beams as a source of illumination. The extremely short wavelength of electron beams contributes to the theoretically high resolution of EM. When a specimen is illuminated by high-energy electrons, various interactions between the specimen and the incident electrons occur, as shown in Figure 7.6. If the specimen is extremely thin, many electrons penetrate it. These electrons are called transmitted electrons or scattered electrons. These are detected using transmission electron microscopy (TEM). Back-scattered electrons and secondary electrons are detected using scanning electron

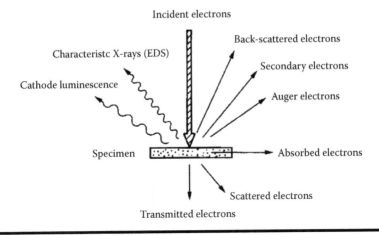

Figure 7.6 **Various waves generated owing to the electron–matter interaction.**

microscopy (SEM) (discussed in Section 7.4). Characteristic x-rays resulting from inner shell excitations are detected using energy dispersive spectrometry (EDS), which is used for qualitative and quantitative elemental analysis of TEM and SEM. Characteristic x-rays for x-ray photoelectron spectroscopy (XPS) and Auger electrons for Auger electron spectroscopy (AES) will not be discussed in this book.

7.3.1 Principle

7.3.1.1 Electromagnetic Lens

Since glass lenses cannot be used in EM, electromagnetic lenses (electrostatic lenses or magnetic lenses) are used instead to focus electron beams.

If electrons enter a high electric field U_2 from a low electric field U_1, they are deflected, similar to the refraction of light rays by glass lenses, and their velocity changes to v_2 from v_1, as shown in Figure 7.7. If the initial velocity of the electrons is 0, we have:

$$v_1 = \sqrt{\frac{2eU_1}{m^2}}, v_2 = \sqrt{\frac{2eU_2}{m^2}} \tag{7.7}$$

Then:

$$\frac{\sin\theta}{\sin\gamma} = \frac{v_2}{v_1} = \sqrt{\frac{U_2}{U_1}} = \frac{\lambda_1}{\lambda_2} \tag{7.8}$$

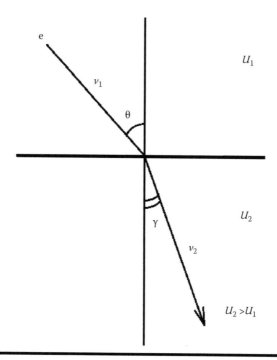

Figure 7.7 Schematic of an electrostatic lens.

Figure 7.8 Schematic of a magnetic lens.

Therefore, a group of isoelectric fields with certain shapes can be used to deflect and focus the electrons. An electrostatic lens is constructed accordingly.

Similarly, electrons with initial velocity are deflected in a magnetic field. Therefore, a magnetic field generated by a cylindrical coil with a current flowing in it will focus electron beams, yielding a magnetic lens, as shown in Figure 7.8.

7.3.1.2 Resolution of Electromagnetic Lens

For an electron with mass m and speed v, the wavelength is:

$$\lambda = \frac{h}{mv} \tag{7.9}$$

Here h is Planck's constant, $h = 6.62 \times 10^{-34}$ J·s. When electrons are accelerated through a potential U, their potential energy can be described as:

$$\frac{1}{2}mv^2 = eU \tag{7.10}$$

Here e is the electron's charge, $e = 1.60 \times 10^{-19}$ C. The electron's wavelength, as a function of the accelerating voltage, is:

$$\lambda = \frac{h}{\sqrt{2eUm}} \tag{7.11}$$

The accelerating voltage in EM is typically very high, implying that the electrons' speed has to be corrected for relativistic effects. The relativistic wavelength of electrons is:

$$\lambda = \frac{1.225}{\sqrt{U(1 + 0.9788 \times 10^{-6}U)}} \, (nm) \tag{7.12}$$

Table 7.2 lists the wavelengths of electrons for various accelerating voltages.

The aperture semi-angle of an electromagnetic lens is typically 10^{-3}–10^{-2} rad. The angle α is quite small; thus, $\sin \alpha \approx \alpha$. Since EM is performed in vacuum, the refractive index is $n = 1$. Thus, the resolution limit due to diffraction, Δr_d in Equation 7.4, becomes:

$$\Delta r_d \approx 0.61\lambda / \alpha \tag{7.13}$$

Therefore, higher accelerating voltages yield shorter wavelengths of electron beams. High resolution can be achieved for short electron wavelengths and large α. For example, when $U = 100$ kV, $\alpha = 10^{-2}$ rad, $\Delta r_d = 0.225$ nm.

The resolving power of an electromagnetic lens depends not only on diffraction, but also on chromatic and spherical aberrations; note that spherical aberration cannot be compensated by introducing diverging lenses as in light optics.

Table 7.2 Wavelengths of Electrons for Various Accelerating Voltages

U/kV	20	30	50	100	200	500	1000
$\lambda/(10^{-3}$ nm)	8.59	6.98	5.36	3.70	2.51	1.42	0.87

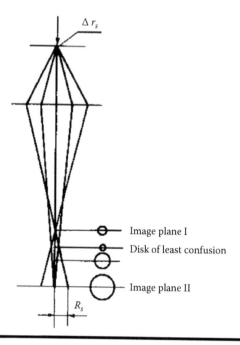

Figure 7.9 **Principle of spherical aberration in an electromagnetic lens.**

The effect of spherical aberration is shown in Figure 7.9. The off-axis magnetic field reflects electrons more than the paraxial magnetic field; thus, electrons with larger α will be focused earlier than those with smaller α. As a result, point objects are imaged as finite-size disks, which limits the resolution of EM. If R_s is the radius of the disk of the least confusion caused by spherical aberration, the resolution limit owing to the spherical aberration, Δr_s, at the object, will be:

$$\Delta r_s = R_s / M = C_s \alpha^3 \tag{7.14}$$

Here, C_s is the coefficient of spherical aberration.

Therefore, the theoretical resolution of an electromagnetic lens, Δr, is approximately the sum of the effects of spherical aberration Δr_s and diffraction Δr_d.

$$\Delta r = \Delta r_s + \Delta r_d = C_s \alpha^3 + 0.61 \lambda / \alpha \tag{7.15}$$

The theoretical resolution Δr changes with α, as shown in Figure 7.10. The minimal Δr_{min} is the theoretical resolution limit, obtained for the optimal aperture semi-angle α_{opt}.

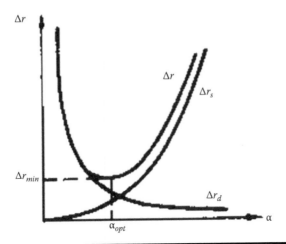

Figure 7.10 **Relation between the resolution of an electromagnetic lens and aperture semi-angle.**

7.3.1.3 Depth of Focus and Depth of Field

Typically, small α is used to minimize spherical aberration. At the same time, a better depth of field (D_f) and depth of focus (D_l) are obtained as well.

All specimens have a certain thickness associated with them. Only the points in the ideal object plane can be focused in the image plane, points in other planes form an out-of-focus disk. If the size of the disk is no more than that caused by spherical aberration and diffraction, the resolving power is not affected. The possible axial deviation from the object plane is defined as the depth of field D_f, as shown in Figure 7.11.

$$D_f = \frac{2\Delta r}{tg\alpha} \approx \frac{2\Delta r}{\alpha} \tag{7.16}$$

Typically, α is 10^{-3}–10^{-2} rad, so the depth of field is $(200$–$2000)\Delta r$. For a TEM with an accelerating voltage of 100 kV, if the thickness of a specimen is no more than 200 nm, all parts of the specimen can be imaged clearly. This is a significant advantage over light microscopy, especially for high magnifications.

If the detector is not in the ideal image plane, out of focus imaging occurs. Similarly, if the size of the out-of-focus disk is not larger than that of the disk owing to spherical aberration and diffraction, the resolving power is not affected. The possible axial deviation from the image plane is defined as the depth of focus D_l, as shown in Figure 7.12.

$$D_l = \frac{2\Delta r}{\alpha} M^2 \tag{7.17}$$

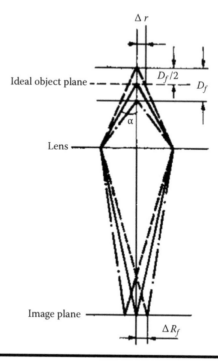

Figure 7.11   Depth of field of an electromagnetic lens.

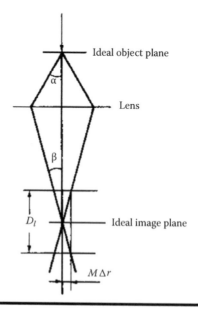

Figure 7.12   Depth of focus of an electromagnetic lens.

Thus, a TEM also has a very large depth of focus. That is to say, even if the image plane deviates significantly from the ideal image plane, the image can be sufficiently clear. Considering that there are many lenses, and the total magnification is the product of magnifications of all lenses, the depth of focus can reach 10–20 cm. This is also very much advantageous for microscopic observations.

7.3.1.4 Diffraction Contrast

The contrast of an electron image is the difference between electron intensities across different regions, i.e., light and shade contrast. The contrast, depending on the scattering, absorption, interference, and diffraction of illuminating electrons by a specimen, can be divided into three categories:

■ Scattering Contrast

The scattering contrast, also known as the mass-thickness contrast, is based on the imaging of electrons scattered by atoms in a specimen. As shown in Figure 7.13, the incident electrons are scattered at all angles. The scattering intensity decreases with increasing the scattering angle, so it is the strongest in the "forward" direction. Electrons that are scattered at large angles are blocked by the aperture, and those that are scattered at small angles or transmitted reach the imaging plane (the detector). In a specimen, domains with higher thickness, higher density, or higher atomic number scatter more electrons, and thus appear darker. The scattering contrast contributes to the main contrast of an amorphous specimen, or to amorphous regions in polymer materials.

■ Diffraction Contrast

For a thin crystallized polymer specimen with uniform thickness and density, a good electron micrograph cannot be attributed to the scattering

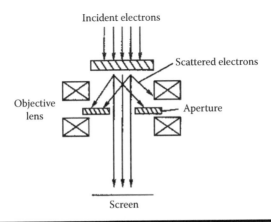

Figure 7.13   Schematic of the scattering contrast generation.

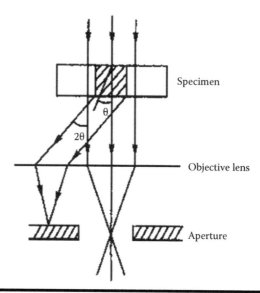

Figure 7.14   Schematic of the diffraction contrast generation in a light-field image.

contrast, but rather to the diffraction contrast, which reflects the diffraction of electrons by the crystal. As shown in Figure 7.14, if the diffracted beam is blocked by an aperture and the transmitted beam reaches a screen, a *bright-field image* is obtained. If the image is observed using the diffracted beam, the image is called a *dark-field image* (not discussed here).

■ Phase Contrast

For a very thin specimen (< 60 nm), neither the scattering contrast nor the diffraction contrast contribute to electron micrographs; rather, the phase contrast becomes important. When electrons pass through a specimen, their energy and wavelength change somewhat, leading to optical path differences and phase differences. The interference of electron beams with different phases gives rise to the phase contrast.

7.3.2 Instrument and Specimen Preparation

7.3.2.1 Structure of TEM

Similar to optical microscopes, a typical TEM comprises an illumination system, a specimen chamber, an imaging system, a viewing and recording system, and a vacuum system. The illumination system comprises an electron gun to generate the beam of electrons, and condenser lenses for focusing the electrons. The specimen chamber contains a specimen holder and a specimen stage, in which the holder allows for the tilting, rotation, heating, cooling, and straining of the specimen. The imaging system is used to obtain the morphology and the diffraction pattern of the

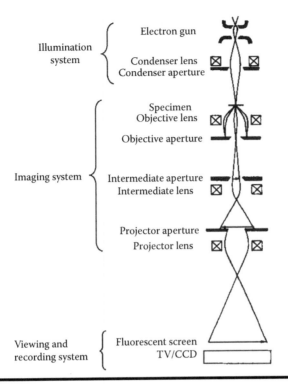

Illumination system
- Electron gun
- Condenser lens
- Condenser aperture

Imaging system
- Specimen
- Objective lens
- Objective aperture
- Intermediate aperture
- Intermediate lens
- Projector aperture
- Projector lens

Viewing and recording system
- Fluorescent screen
- TV/CCD

Figure 7.15   Schematic diagram of the TEM structure.

specimen through multiple magnifications (i.e., objective, intermediate), as well as through projection lenses and apertures. Images are viewed on a screen, or are recorded on a film or using a charge-coupled device (CCD) camera. The schematic diagram is shown in Figure 7.15.

To avoid the interference effects due to air, with air molecules scattering and absorbing electrons and interfering with the specimen imaging, a high vacuum system (typically 10^{-5}–10^{-4} Pa) is necessary to protect the electron gun from degradation and maintain the stability of the corresponding TEM.

7.3.2.2 Specimen Preparation

Specimen preparation significantly affects the quality of imaging in TEM. The specimen must be sufficiently thin to allow electrons to penetrate; its thickness or diameter are typically 100–200 nm. Such a thin film or fine powder cannot support itself; consequently, a grid with a support film on it is employed (typically a carbon film with thickness not exceeding 20 nm). The typical grid diameter is 3 mm; thus, the specimen must be sufficiently small to fit into the grid. Since the specimen only represents a very small portion of a sample, selecting a representative specimen is an important issue.

Various preparation techniques have been developed for different types of specimens. Here, we describe some common techniques for preparation of polymer specimens. Other techniques, such as surface replica and ion-beam thinning, are used for preparing metallic and ceramic specimens, and will not be discussed here.

For emulsion or suspension samples, one can dilute the sample and drip drops of liquid on a grid with a supporting film, and then dry the sample to prepare it for TEM observations. In some cases, sonication is applied to achieve good dispersion of particles in the emulsion or suspension before dripping.

For multicomponent, multiphase samples, an ultramicrotome is often used to cut sections that are sufficiently thin (thickness, 50–200 nm) for TEM observations. Elastic samples, e.g., rubbers, cannot be cut at room temperature; thus, they are often cut at liquid nitrogen temperature, known as cryo-ultramicrotome. Fragile or flexible samples are often embedded in cured media, such as epoxy resins and polyacrylates, before cutting. The composition of a curable medium is adjusted to adapt to the rigidity of the processed sample.

Pure polymers demonstrate low contrast in TEM, owing to their similar elemental compositions; consequently, etching and staining techniques are often used. Etching amounts to selectively removing some parts from a specimen using a solvent or strong oxidant. Staining amounts to treating a specimen with agents that contain heavy metal elements. Some parts of the treated specimen absorb or interact with metal elements, thus increasing the ability to scatter incident electrons. Typical staining agents include OsO_4 or RuO_4, which crosslink and stain unsaturated double bonds in the rubber phase.

7.3.3 Applications in Polymer Materials

TEM is an important tool for polymer research, especially for investigating polymer/inorganic composites and crystallization in polymers. TEM has an extremely high resolution (under 0.2 nm), which implies that very small regions can be observed at a very large magnification. Considering that heterogeneous structures are very common in polymer materials, a representative specimen must be prepared and representative micrographs must be reported. One good way of doing this is to observe a specimen at a low resolution, and then photograph the desired representative region at a high resolution.

7.3.3.1 Nanocomposites

In nanocomposites, nanofiller particles are expected to be dispersed evenly in the polymer matrix, which contributes to a high performance of nanocomposites. The dispersion status of nanofiller particles can be easily observed using TEM. For example, montmorillonite (MMT) is a typical layered silicate nanofiller, in which many layers (thinner than 1 nm) are stacked together to form microsize particles. An organic ammonium salt enters MMT layers to increase the lamellar spacing. Then, monomer molecules diffuse into the MMT and polymerize to prepare a nanocomposite. Figure 7.16 shows a TEM micrograph of a PA/MMT nanocomposite obtained by *in situ* polymerization. The black fibers are the MMT layers in the PA matrix. Most of the MMT are exfoliated, shown as isolated layers, but there are still some intercalated MMT, shown as stacked layers.

Recently, carbon nanofillers such as carbon nanotubes and graphene have attracted significant attention owing to their extremely high strength, electrical, and thermal conductivities. The fine structures of these materials can only be observed using TEM. For example, Figure 7.17 shows the fine structure of a

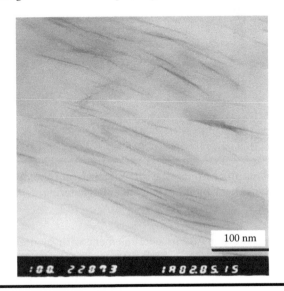

Figure 7.16 TEM micrograph of a PA/MMT nanocomposite. (Courtesy of Jian Yu, Department of Chemical Engineering, Tsinghua University.)

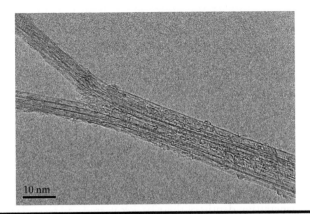

Figure 7.17 Fine structure of a multiwall carbon nanotube and polymer chains on it. (Courtesy of Ling Hu, Department of Chemical Engineering, Tsinghua University.)

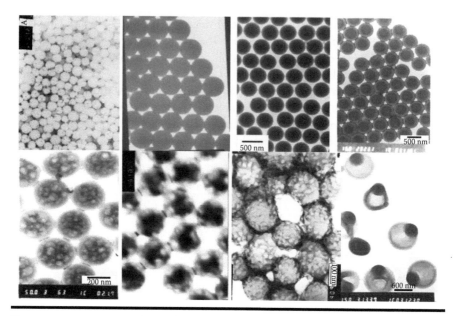

Figure 7.18   Various microspheres with different morphologies. (Courtesy of Prof. Chengyou Kan, Department of Chemical Engineering, Tsinghua University.)

multiwall carbon nanotube and its entanglement with monodispersed polymer chains.

7.3.3.2 Microspheres and Microcapsules

Microspheres and microcapsules are usually observed in emulsions. Figure 7.18 shows a variety of microspheres: uniform-size, core-shell, porous, strawberry-like, and Janus-like. Figure 7.19 shows various hybrid microcapsules prepared at different polymer concentrations. At high concentrations, sunflower-like hybrid microcapsules are obtained [1].

In addition, crystal lattice structures of polymers can be studied by analyzing diffraction patterns of transmitted electron beams.

7.4 Scanning Electron Microscopy (SEM)

When a beam of electrons scans the surface of a specimen, the resulting backscattered electrons and secondary electrons, especially the latter, are detected in SEM imaging. The penetration depths of these backscattered and secondary electrons are several microns and 5–50 nm, respectively. SEM is a very convenient and powerful tool for surface morphology observations, with high resolution, high magnification

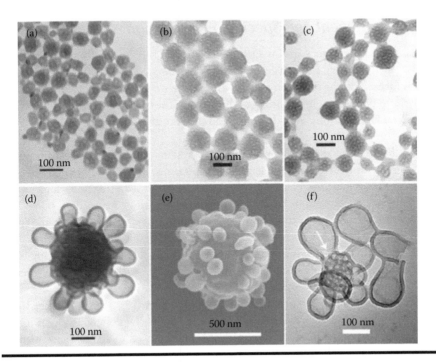

Figure 7.19 Hybrid microcapsules at different polymer concentrations. (From http://www.iccas.ac.cn)

range (from 30 to 3×10^5 x), high contrast, high depth of field, high depth of focus, and three-dimensional aspect. Strikingly, SEM images appear as though they are illuminated by a light beam from above, as shown in Figure 7.20. Owing to these numerous advantages, this method has become the most widely used one for microstructural analysis.

Figure 7.20 Environment scanning micrographs of a compound eye (a) and an insect's antenna (b). (Courtesy of Wenyan Yang, State Key Laboratory of Tribology, Tsinghua University.)

7.4.1 Instrument and Specimen Preparation

A SEM comprises four parts, as shown in Figure 7.21. A beam of high-energy electrons, generated by an electron gun, passes through two to three electromagnetic lenses and is focused on the specimen surface. The scan system drives the beam of electrons to perform point-by-point scanning of the surface. The excited backscattered electrons and secondary electrons enter detectors, and then are amplified and output.

■ Illumination system: The illumination system comprises an electron gun, a group of electromagnetic lenses and apertures, and a specimen chamber. It produces a beam of electrons that is focused and deflected at various angles, so that various points on the specimen surface can be scanned.
■ Scan system: The scan system drives the beam of electrons to scan the specimen surface at different speeds and directions.
■ Detecting and viewing system: Various detectors are used for detecting backscattered electrons, secondary electrons, and x-rays. The signals are received and amplified for on-screen imaging.
■ Vacuum system: As in a TEM, a vacuum system (10^{-5}–10^{-6} Pa) is needed in a typical SEM to protect the electron gun. By special design, an environment scanning electron microscope (ESEM) can be used to analyze specimens in low vacuum (10^{1}–10^{2} Pa).

Resolution, magnification, and contrast are three key parameters in SEM.

The resolution of SEM depends on the diameter of the electron beam. In advanced field emission SEM (FESEM), the resolution can be as high as 10^{-1} nm.

The magnification of SEM is defined as the ratio of the screen size to the scanned range of a specimen. The smaller the scanned range, the higher the magnification.

The contrast of SEM is related to the surface morphology, the atomic number, and the voltage.

The secondary electron yield reflects the surface morphology of a specimen. On a rough surface, the beam of incident electrons illuminates various positions

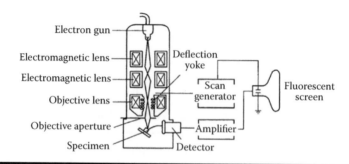

Figure 7.21 Schematic of a SEM.

at different angles to yield different amounts of secondary electrons, thus yielding different signal intensities and brightness. Larger angles of incidence lead to higher secondary electron yields. Sharp edges, small particles, and pit edges contribute significantly to the yield of secondary electrons. The so-formed surface morphology contrast is shown in Figure 7.22.

The yield of secondary electrons also depends on the atomic number of a specimen. Elements with high atomic numbers yield more secondary electrons. For polymers, because the elemental compositions are similar, the atomic number contrast is low. However, if inorganic fillers are present, the contrast is high owing to the presence of heavy elements.

If the specimen surface has voltage gradients, a larger number of secondary electrons will be emitted from regions with higher voltage compared with those emitted from lower voltage regions, giving rise to the voltage contrast. Polymer materials are generally not conductive; consequently, a local accumulation of charge can easily occur, resulting in a very high voltage contrast, which sometimes damages the analyzed specimen. Therefore, a thin metal layer (such as a ~10-nm-thick layer of gold) is sprayed on the specimen surface before performing SEM observations. Figure 7.23 shows micrographs of a fiber sample. Figure 7.23a shows a typical image

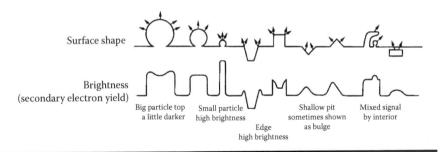

Figure 7.22   Schematic diagram of the surface morphology contrast.

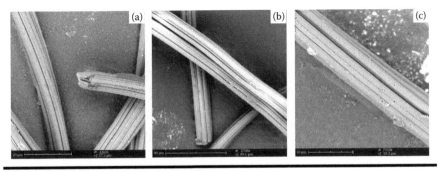

Figure 7.23   SEM images of a bundle of fibers. (a) Normal case. (b) Case with surface charge accumulation. (c) Breakdown of the surface owing to the accumulation of charge.

obtained after applying a gold spray on the specimen surface. Figure 7.23b shows that a very bright image is obtained owing to the local accumulation of charge for a specimen imaged without a gold spray. Figure 7.23c shows holes on the fiber surface induced by a large amount of accumulated charge.

Specimens can be easily prepared for SEM observations. A conductive or semi-conductive specimen is fixed on the specimen stage using a conductive adhesive. For nonconductive specimens, such as polymer materials, a metal layer (typically tens of nanometers thick) is sprayed on the specimen surface before observations, in order to eliminate charge accumulation and prevent accumulation-induced break-down of the specimen. The specimen chamber can accommodate specimens with the typical size of several centimeters, and a specimen can be rotated or moved in the x, y, and z directions. The analyzed specimen must be stable with respect to the irradiation by electron beams; otherwise, melting or decomposition may occur. The analyzed specimen must not contain volatile components, or the chamber may be polluted.

7.4.2 Applications in Polymer Materials

SEM is one of the most widely used analytical instruments. It generates a wealth of information related to the surface morphologies of a variety of specimens.

Pore morphology and size of membrane for separation can be observed using SEM. Examples for a lithium-ion battery separator and a fuel cell separator are shown in Figures 7.24 and 7.25, respectively.

SEM is often used to observe phase morphologies in multiphase, multicompo-nent materials. Figure 7.26 shows the cross section of a PS/PA6 alloy [2]. As PA6 increases, the phase morphology changes from that of dispersed particles to that of the bicontinuous phase, and finally to that of the continuous phase. The phase size changes accordingly.

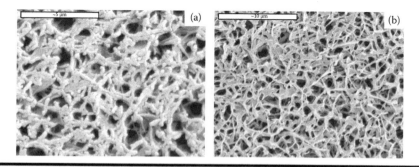

Figure 7.24 P(VDF-HFP) membrane used in a lithium-ion battery (average pore size is 0.6 µm). (a) Cross-section view and (b) surface view. (Courtesy of Prof. Xiao Zhou, Department of Chemical Engineering, Tsinghua University.)

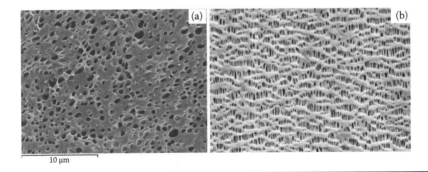

Figure 7.25 PP membranes prepared using (a) the thermally induced phase separation (TIPS) method (a) and the stretching method (b), used in fuel cells. (Courtesy of Prof. Baohua Guo, Department of Chemical Engineering, Tsinghua University.)

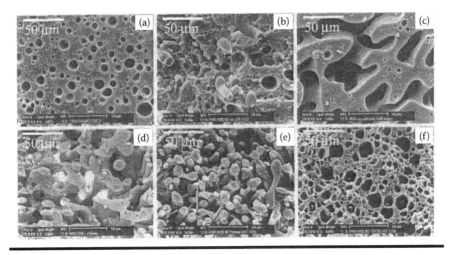

Figure 7.26 Cross section of a PS/PA6 alloy. The content of PA6 increases gradually through a–f. In a–e, PA6 was etched using formic acid. In f, PS was etched using chloroform. (From Tol, R.T. et al., *Polymer*, 2004, 45(8), 2587–2601.)

Microstructure and nanostructure are well observed in SEM. Figure 7.27a shows the photograph of a *gerris remigis* walking on water. This insect can remain on the water surface owing to the micro- and nanostructure of setae on its legs, the SEM images of which are shown in Figure 7.27b and c [3]. This specific structure is hydrophobic and endows a *gerris remigis* the talented ability to freely walk on the water surface. Inspired by the leg structure of *gerris remigis* and other biologic structures, researchers have developed various novel biomimetic materials with extraordinary properties, such as super hydrophilic materials, super hydrophobic materials, super adhesive materials, and materials that exhibit special optical effects.

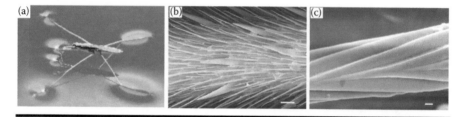

Figure 7.27 **(a) Photograph of a gerris remigis on the water surface. (b) SEM image of setae on the leg of a gerris remigis (the scale is 20 μm). (c) Nanoscale spiral grooves on a seta (the scale is 200 nm). (From Gao, X., and L. Jiang. *Nature*, 2004, 432(7013), 36.)**

7.5 Atomic Force Microscopy

Atomic force microscopy (AFM) principally differs from optical and electron microscopies. AFM is derived from scanning tunneling microscopy (STM) based on the quantum tunneling effect. It allows extremely high resolution surface observations under normal conditions.

7.5.1 Principle

In AFM, a surface is scanned using a sharp probe tip. The minute forces between the scanned material's atoms and the tip are detected and controlled to measure the surface contours on the atomic level, which explains the name AFM.

Figure 7.28 schematically shows the main components of a typical AFM system. A sample is placed on a stage that can be moved in the x, y, and z directions. One end of a cantilever is fixed, and the tip at another end approaches or contacts the sample. A laser beam is bounced from a mirror mounted on the back of the probe, to a detector. When the tip is quite close to the specimen surface, the minute atomic forces attract or reject the tip, so the cantilever is deflected and causes the laser beam to shift on the detector. Therefore, by measuring the shift, the surface contours of a sample or the minute interaction forces are detected.

There are three measurement modes in AFM: the contact mode, the noncontact mode, and the tapping mode.

In the contact mode measurement, the probe tip does not really contact the specimen surface. The distance is under 0.03 nm, so there is a repulsion between the tip atoms and the specimen surface atoms. In the contact mode, measurements can be performed in two ways, as shown in Figure 7.29. One way is to control the distance and repulsion constant, so the tip moves up and down with the surface contours. Another approach is to keep the tip height constant, so the repulsion changes with the tip-surface distance. Contact mode measurements are typically performed on rigid surfaces with atomic-level resolution.

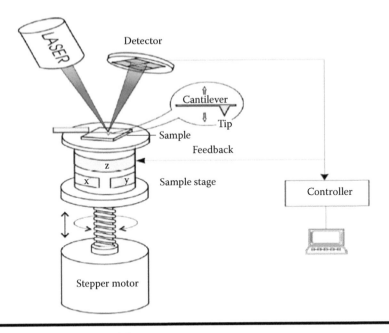

Figure 7.28   Schematic of the AFM system.

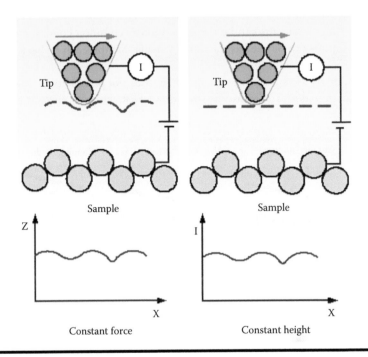

Figure 7.29   Two measurement methods in the contact mode of AFM.

For soft samples, contact mode measurement often damages the specimen surface and causes pollution of the tip; consequently, measurements are often performed in the noncontact mode. The distance between the tip and the specimen surface is on the order of 10^2 nm. The attractive interaction changes the frequency and amplitude of inherent vibration modes. The signal is relatively weak, with low resolution.

In order to overcome the shortcomings of noncontact mode measurements, measurements are often performed in the tapping mode for polymer materials. The tip vibrates with an amplitude of ~100 nm; thus, it frequently contacts the measured specimen surface. The contact time is so short that there is no damage or pollution, and high resolution can be achieved.

For comparison, the surface contours of the same polymer material, obtained using the three above-described measurement modes, are shown in Figure 7.30. In the contact mode, the specimen surface is easily hooked up, so a false "hump" appears; in the noncontact mode, the resolution is relatively low; and in the tapping mode, a good result is obtained.

AFM has many advantages:

■ Both conductive and nonconductive samples can be measured.
■ A three-dimensional image of the specimen surface can be obtained with atomic or subatomic resolutions. The vertical and horizontal resolutions are ~0.01 nm and ~0.1 nm, respectively. The resolution mainly depends on the tip shape.
■ AFM measurements can be performed under common conditions. Both air and liquids can be used as the sample media. Samples can be heated/cooled.

Although AFM is a sensitive tool, AFM does not satisfactorily resolve steep walls; thus, it requires a relatively smooth surface, i.e., the surface fluctuation

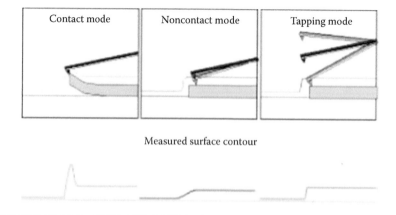

Measured surface contour

Figure 7.30 Surface contours obtained in the three measurement modes.

amplitude must be no more than hundreds of nanometers. AFM can scan only in a limited range, and the scanning process is slow.

7.5.2 Instrument and Sample

A typical AFM system is composed of three parts:

■ Detection System

  The detection system contains a probe, a laser, and a photo detector. The probe includes a cantilever and a tip. The cantilever is 100–200 μm long, with a mirror mounted on the back for bouncing the laser beam. For contact mode measurements, Si or Si_3N_4 cantilevers are used, with the elasticity coefficient in the 10^{-2}–10^2 N/m range. The resonance frequency is above 10 kHz. For tapping mode measurements, the cantilever is made of a single crystal of Si, with the elasticity coefficient in the 20–100 N/m range. The resonance frequency is 200–400 kHz. The tip is pyramidal or conical, with a curvature radius of several to tens of nanometers.

■ Scanning System

  A stepper motor drives the sample stage to move in the x, y, and z directions to implement surface scanning by the tip.

■ Feedback System

  By *in situ* measuring the interaction force, a computer-controlled feedback system maintains a constant distance between the specimen surface and the probe tip, maintaining the tip at a constant height above the surface.

7.5.3 Applications in Polymer Materials

AFM is widely used for resolving fine structures on surfaces, for surface nanoprocessing, and for minute force determination, to name a few applications.

Since its introduction, AFM has proven to be a powerful tool for nanoprocessing. As shown in Figure 7.31, atoms can be positioned using AFM to form various patterns [4,5].

Using AFM, fine structures on the specimen surface can be easily observed; for example, surface relief gratings and microphase separation morphology (Figure 7.32).

AFM has been also used for *in situ* observations of crystallization, under various conditions. Figure 7.33 shows the AFM images during the crystallization process of poly(hydroxybutyrate-co-hydroxyhexanoate). The images capture the growth process of a banded spherulite, from the melt. The right column shows the corresponding height profiles. The peak-valley difference for this banded spherulite did not change during the crystallization process.

AFM utilizes the atomic force between the tip and the specimen surface; thus, it is a powerful method for detecting the force of a single molecule to obtain modulus. As shown in Figure 7.34, two strands from a DNA helix were fixed on a substrate

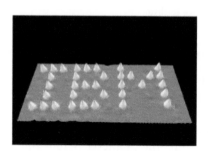
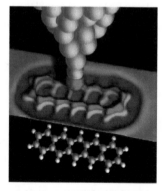

Figure 7.31 AFM for nanoprocessing. (From http://www-03.ibm.com/press/us /en/photo/28500.wss; and Gross, L. et al., *Science*, 2009, 325 (5944): 1110.)

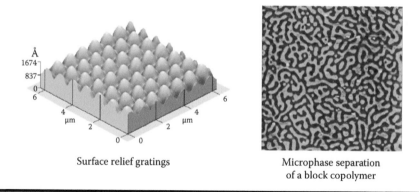

Surface relief gratings

Microphase separation of a block copolymer

Figure 7.32 AFM images of surface relief gratings and microphase separation of a block copolymer. (Courtesy of Prof. Xiaogong Wang and Prof. Xuming Xie, Department of Chemical Engineering, Tsinghua University.)

and on the tip, respectively. The two strands were allowed to approach and assemble through hydrogen bonds into a helix, and then, the substrate was moved away from the tip until the helix separated into two strands. The force and the displacement during the procedure were recorded and are shown in Figure 7.34. From these data, the rupture force for hydrogen bonds can be obtained.

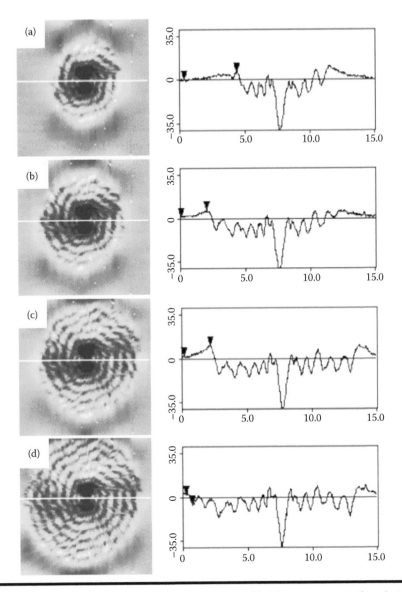

Figure 7.33 ATM observations of the crystallization process of poly(R-3-hydroxybutyrate-co-R-3-hydroxyhexanoate. Crystallization temperature: 45°C, tapping mode, time steps of 17 min. (Courtesy of Prof. Jun Xu, Department of Chemical Engineering, Tsinghua University.)

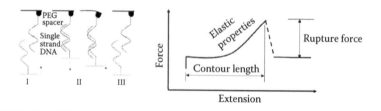

Figure 7.34 **Hydrogen bond force between two strands of DNA, determined using AFM. (Courtesy of Prof. Wenke Zhang, State Key Laboratory of Supramolecular structure and Materials, Jilin University.)**

References

1. http://www.iccas.ac.cn
2. Tol, R.T., G. Groeninckx, I. Vinckier, P. Moldenaers, and J. Mewis. Phase morphology and stability of co-continuous (PPE/PS)/PA6 and PS/PA6 blends: Effect of rheology and reactive compatibilization. *Polymer*, 2004, 45 (8): 2587–2601.
3. Gao, X., and L. Jiang. Biophysics: Water-repellent legs of water striders. *Nature*, 2004, 432(7013), 36.
4. http://www-03.ibm.com/press/us/en/photo/28500.wss
5. Gross, L., F. Mohn, N. Moll, P. Liljeroth, and G. Meyer. The chemical structure of a molecule resolved by atomic force microscopy. *Science*, 2009, 325 (5944): 1110.

Exercises

1. Theoretical resolution of an electromagnetic lens depends on spherical aberration and diffraction effects. Derive the expressions for the optical aperture semi-angle α_{opt} and the resolution limit Δr_{min}.
2. For an electromagnetic lens with the aperture semi-angle in the 10^{-2}–10^{-3} rad range, resolution of 1 nm, and magnification of 1000×, calculate the depth of field D_f and the depth of focus D_l.
3. Summarize the advantages and disadvantages of TEM.
4. Summarize the advantages and disadvantages of SEM.
5. Is it possible to "see" chemical bonds using AFM? Why or why not?

Chapter 8

X-Ray Diffraction

8.1 Geometry of Crystals

X-ray diffraction is mainly used for investigations of crystal structures. Therefore, the fundamental geometry and structure of crystals are introduced first, explaining how the atoms are arranged in the crystals and how crystals diffract x-rays.

8.1.1 Lattice and Crystal System

A crystal is a solid composed of structural units (imaginary points) arranged in a pattern that is periodic along the three dimensions. These imaginary points constitute a point lattice, which is defined as an array of points in space, and arranged so that each point has identical surroundings [1]. Each unit cell, as the basic unit of a crystal, is constituted by three sets of planes to a parallelepiped, as shown in Figure 8.1. The size and shape can be described by the three lengths a, b, c and the angles between them (α, β, γ). These lengths and angles are the lattice parameters of the unit cell. Based on the symmetries and shapes of unit cells, all crystals are classified into seven crystal systems. Furthermore, according to the point arrangements (at the corners, at the center of a unit cell, or in the center of the unit cell's face), only 14 point lattices (Bravais lattices) are possible; these are listed in Table 8.1.

8.1.2 Plane Index

A crystal plane is tilted with respect to the crystal axes (*a*, *b*, *c*) and intercepts the three axes. The reciprocals of the intercepts can be calculated and converted to a set of smallest integers by multiplication or division throughout. We thus obtain

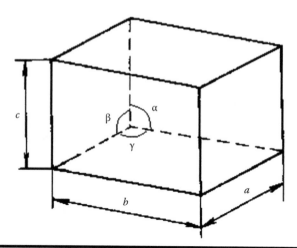

Figure 8.1   Unit cell.

a symbolism for the orientation of a plane in a lattice, the *plane indices* (*hkl*), written in parentheses (also known as *Miller indices*). For example in Figure 8.2, if the axial unit lengths are *a*, *b*, *c*, the plane makes the intercepts of 1/2, 1/3, 2/3. The reciprocals are 2, 3, 3/2. The corresponding smallest integers are 4, 6, 3. Thus, the plane indices are (463). This usually refers to the plane which is nearest to the origin *O*, but can also refer to any other parallel equidistant plane or to the whole set of planes taken together.

When a plane is parallel to a certain axis, the intercept is infinity, and the corresponding index is 0. If a plane cuts a negative axis, the corresponding index is negative and is written with a bar over it, e.g., $(00\bar{1})$.

A slightly different system of plane indexing is used in the hexagonal system. There are four axes a_1, a_2, a_3, c, and so the plane indices are written (*hkil*). The index *i* depends on the values of *h* and *k*. The relation is:

$$i = -(h+k) \tag{8.1}$$

For example, the plane indices of top (in blue), right side (in grey), left side (in light red), and front (in red) planes of the hexagonal crystal in Figure 8.3 are (0001), $(11\bar{2}0)$, $(\bar{1}\bar{1}00)$, and $(10\bar{1}0)$.

8.1.3 *Interplanar Spacing*

In Figure 8.4, the origin *O* is in one plane in the (*hkl*) set of planes. The nearest parallel plane intercepts the axes at *A*, *B*, *C*. The plane normal passes through the origin *O*, and the distance *ON* is the interplanar spacing d_{hkl}.

Table 8.1   Crystal Systems and Bravais Lattices

System	Axial lengths and angles	Bravais lattice			
		Simple (P)	Body-centered (I)	Base-centered (C)	Face-centered (F)
Cubic	$a = b = c$ $\alpha = \beta = \gamma = 90°$				
Tetragonal	$a = b \neq c$ $\alpha = \beta = \gamma = 90°$				
Orthorhombic	$a \neq b \neq c$ $\alpha = \beta = \gamma = 90°$				

(Continued)

Table 8.1 (Continued)   Crystal Systems and Bravais Lattices

System	Axial lengths and angles	Bravais lattice			
		Simple (P)	Body-centered (I)	Base-centered (C)	Face-centered (F)
Trigonal	$a = b = c$ $\alpha = \beta = \gamma \neq 90°$				
Hexagonal	$a = b \neq c$ $\alpha = \beta = 90°$ $\gamma = 120°$				

(Continued)

Table 8.1 (Continued)   Crystal Systems and Bravais Lattices

System	Axial lengths and angles	Bravais lattice			
		Simple (P)	Body-centered (I)	Base-centered (C)	Face-centered (F)
Monoclinic	$a \neq b \neq c$ $\alpha = \gamma = 90° \neq \beta$				
Triclinic	$a \neq b \neq c$ $\alpha \neq \beta \neq \gamma \neq 90°$				

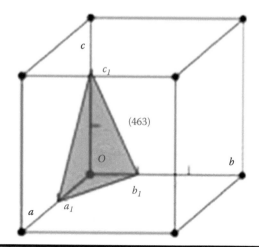

Figure 8.2   Plane indices.

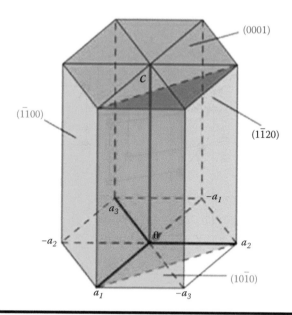

Figure 8.3   Plane indices in the hexagonal system.

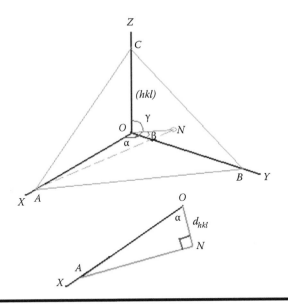

Figure 8.4 Interplanar spacing.

Thus:

$$\cos \alpha = ON \,/\, OA = d \,/\, OA = d \,/\, ma$$
$$\cos \beta = ON \,/\, OB = d \,/\, nb \qquad\qquad (8.2)$$
$$\cos \gamma = ON \,/\, OC = d \,/\, pc$$

m, n, p are the reciprocals of $h, k, l,$ and thus:

$$\cos \alpha = d \,/\, (a \,/\, h)$$
$$\cos \beta = d \,/\, (b \,/\, k) \qquad\qquad (8.3)$$
$$\cos \gamma = d \,/\, (c \,/\, l)$$

In the orthorhombic system, $\cos^2 \alpha + \cos^2 \beta + \cos^2 \gamma = 1$, the interplanar spacing equation is:

$$d_{hkl} = \frac{1}{\sqrt{h^2 \,/\, a^2 + k^2 \,/\, b^2 + l^2 \,/\, c^2}} \qquad\qquad (8.4)$$

In the tetragonal system, $a = b$, the interplanar spacing equation is:

$$d_{hkl} = \frac{1}{\sqrt{(h^2 + k^2)/a^2 + l^2/c^2}} \qquad (8.5)$$

The interplanar spacing equation of the cubic system takes on a relatively simple form since $a = b = c$:

$$d_{hkl} = \frac{a}{\sqrt{h^2 + k^2 + l^2}} \qquad (8.6)$$

For the hexagonal system, the interplanar spacing equation is:

$$d_{hkl} = \frac{1}{\sqrt{4(h^2 + hk + k^2)/3a^2 + l^2/c^2}} \qquad (8.7)$$

8.2 Properties of X-Rays

X-rays are short-wavelength electromagnetic radiation (10^{-6}–10 nm), with diffraction x-rays in the 0.01–2.5 nm range. Unit cells of polymer crystals are ~0.2 nm; thus, x-ray diffraction can be used to investigate fine crystal structures of polymers, although this approach has also been used in metallurgy for a long time.

8.2.1 Generation of X-Rays

X-rays are generated when electrons are rapidly decelerated. In an x-ray tube, the electrons from the cathode C are accelerated by the high voltage U and strike the anode A, or the target, with very high speed. X-rays are generated at the point of impact and radiate in all directions, as shown in Figure 8.5.

Continuous Spectrum
X-rays from the metal target A consist of different wavelengths. Some electrons hit the target and lose all of their energy in one impact, giving rise to maximal-energy x-rays (i.e., minimal-wavelength x-rays); this limit is called the short-wavelength limit (λ_{SWL}), as shown in Figure 8.6. Such electrons transfer all of their energy eV to x-rays, so

$$\lambda_{SWL} = \frac{hc}{eV} \qquad (8.8)$$

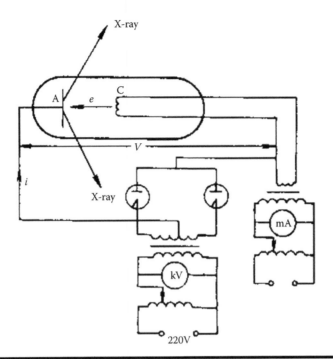

Figure 8.5   Generation of x-rays in an x-ray tube.

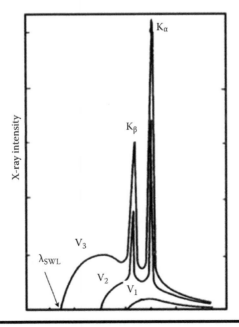

Figure 8.6   X-ray continuous spectrum and characteristic spectrum $(V_3>V_2>V_1)$.

Here, λ_{SWL} is the short-wavelength limit and V is the x-ray tube voltage. High tube voltage V gives rise to short λ_{SWL}.

Other electrons strike the target and successively lose a fraction of their energy after multiple hits. Such electrons give rise to x-rays for which the wavelengths are much longer than λ_{SWL}. All of these wavelengths range upward from λ_{SWL}, constituting a continuous spectrum.

Characteristic Spectrum

In addition to the continuous spectrum, sharp intensities appear at certain wavelengths, called the characteristic spectrum. The characteristic spectrum relates to the atoms of the metal target itself. When one electron strikes the target, a fraction of its energy is transferred to the metal's atom, knocking an atomic electron out of the K shell, and exciting the atom to a higher-energy state. One of the outer electrons immediately exhibits a downward (radiative) transition and fills the vacancy in the K shell, emitting extra energy in the form of an x-ray, and the atom returns to its ground state, as shown in Figure 8.7. The emitted x-ray is called the K radiation. Similarly, L radiation, M radiation, can be obtained. Collectively, this radiation constitutes the characteristic spectrum, in which the energies are the energy differences, i.e., $\Delta E = h\nu$.

The K-shell vacancy may be filled by an electron from any one of the outer shells, thus giving rise to a series of K lines. Transitions from the L or M shells to the K shell are associated with the emission of K_α and K_β lines, respectively, as shown in Figure 8.7. It is more likely for the K-shell vacancy to be filled by an L electron than an M electron; thus, the intensity of the K_α line is stronger than that of the K_β line (Figure 8.6). According to the selection rule, there are two K_α lines, i.e., $K_{\alpha 1}$ and $K_{\alpha 2}$. They are so close to each other that in general they are not resolved and are regarded as simply the K_α line. There are also two K_β lines, but the $K_{\beta 2}$ line is too weak to be resolved. The series of L lines is produced in a similar way.

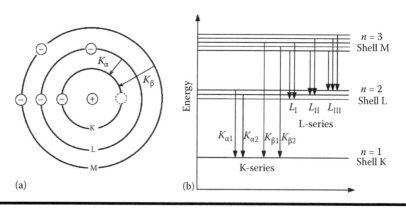

Figure 8.7 **Origin of characteristic x-rays. (a) Atomic electronic transitions. (b) Energy levels, transitions, and corresponding line series.**

The relation between the wavelength of the characteristic x-rays λ and the atomic number Z is:

$$\sqrt{\frac{1}{\lambda}} = K(Z - S) \tag{8.9}$$

where K and S are constants.

Therefore, both the continuous spectrum and the characteristic spectrum of x-rays are produced when electrons strike the metal target. The wavelengths and intensities of characteristic x-rays are related to certain energy differences for specific atoms, and thus can be used for qualitative and quantitative analysis of elements. This is the basis of the x-ray fluorescence analysis (XRF). High-energy electrons knock inner-shell electrons out to free electrons (photoelectrons). By measuring the kinetic energy of these photoelectrons, their binding energies can be calculated. This is the basis of the x-ray photoelectron spectroscopy (XPS). Electrons from outer shells transition downward to fill the vacancies in the inner shells, and the emitted x-rays may excite more outer electrons to free more electrons, which are called the Auger electrons. The characteristic energies of these Auger electrons for various elements are determined in Auger electron spectroscopy (AES). These methods are not discussed here.

8.2.2 Absorption of X-Rays

When x-rays encounter a solid sample, they are partially transmitted and partially absorbed*. Consequently, the intensity of x-rays decreases. The intensity of transmitted x-rays I after passing through a sample of thickness d is:

$$I = I_0 \exp(\mu d) = I_0 \exp(\mu_m \rho d) \tag{8.10}$$

Here, I_0 is the intensity of incident x-ray beams. The coefficients μ and μ_m are the linear absorption coefficient and the mass absorption coefficient, respectively. The parameter ρ is the sample's density.

The mass absorption coefficient of an element, μ_m, varies with wavelength approximately according to the following relation:

$$\mu_m = K\lambda^3 Z^3 \tag{8.11}$$

---

* In fact, a small fraction of x-rays is scattered. Both the scattered and the absorbed x-rays contribute to the reduction in the intensity of the transmitted beam, and thus are said to be absorbed.

Here, K is a constant and Z is the atomic number. Long-wavelength x-rays are therefore easily absorbed, and heavy elements with high atomic number strongly absorb x-rays.

The mass absorption coefficient of a substance containing more than one element, whether it is a mixture or a compound, is a weighted average of the mass absorption coefficients of its constituent elements.

$$\mu_m = W_1\mu_{m1} + W_2\mu_{m2} + W_3\mu_{m3} + \ldots \tag{8.12}$$

Here W_1, W_2, and W_3 are the weight fractions of elements 1, 2, and 3 in the substance, and μ_{m1}, μ_{m2}, and μ_{m3} are the corresponding mass absorption coefficients. Table 8.2 lists the mass absorption coefficients of typical elements in polymers to the Cu K_α line, while the mass absorption coefficients of typical polymers are listed in Table 8.3.

For a specific element, the mass absorption coefficient decreases with increasing x-ray wavelength. When the wavelength decreases to a critical value, i.e., when the energy of the incident x-ray beam increases to a critical value, an inner-shell electron is knocked out to yield a free electron. For a K electron, the critical wavelength is λ_K. For an L electron, the critical wavelength is λ_L. Therefore, at these critical wavelengths, the mass absorption coefficient increases abruptly. As the incident wavelength decreases below the critical value, the absorption coefficient begins to decrease again. Consequently, the curve exhibits some sharp discontinuities that are called absorption edges, as shown in Figure 8.8. There is one K absorption edge λ_K and three L absorption edges λ_{L1}, λ_{L2}, and λ_{L3}.

At the two sides of an absorption edge, the mass absorption coefficient takes on very different values. Based on this feature, monochromatic x-rays can be generated. The beam that emerges from an x-ray tube contains the characteristic K_α and K_β lines, as well as the continuous spectrum. The intensity of undesirable components can be significantly reduced by passing the beam through a filter made of a material whose K absorption edge lies between the K_α and K_β lines of the target metal. For example, in Figure 8.9, the K absorption edge of Ni lies between the K_α and K_β lines of Cu, so Ni can be used as a filter to produce the monochromatic K_α line of Cu.

Table 8.2   Mass Absorption Coefficients of Typical Elements to the Cu K_α Line

Element	H	C	N	O	F	Si	S	Cl
μ_m (cm²/g)	0.435	4.60	7.52	11.5	16.4	60.6	89.1	106

Table 8.3   Mass Absorption Coefficients of Typical Polymers to the Cu K_α Line

Polymer	PE	PP	PAN	PA6	PET	PVA	PTFE	PVC
μ_m (cm²/g)	4.00	4.00	5.13	5.53	6.72	6.72	13.53	61.9

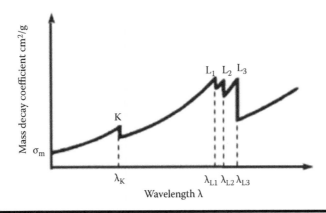

Figure 8.8 **Mass absorption coefficient change vs. wavelength, showing *K* and *L* absorption edges.**

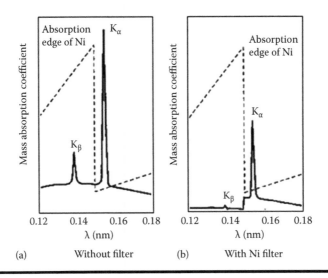

Figure 8.9 **Comparison of Cu spectra (a) before and (b) after passing through a Ni filter. In both plots, the dashed line is the mass absorption coefficient of Ni.**

8.2.3 Scattering and Diffraction of X-Rays

The scattering of x-rays can be categorized as coherent scattering and incoherent scattering.

For inner-shell electrons with strong binding energies, if the energy of x-rays is insufficient to excite these electrons, and elastic collisions between photons in x-rays and electrons only change the direction of the x-rays, no energy loss occurs and the wavelength remains unchanged. The scattered beam has the same wavelength as the incident beam and is said to be coherent with it.

For outer-shell electrons with weak binding energies, if x-rays undergo inelastic collisions with these electrons and transfer energy to them, the electrons will be excited to recoil electrons. Accordingly, the direction of the scattered beam is changed, and the wavelength is increased. Consequently, the scattered beam is not coherent with the incident beam and appears as the background noise in the XRD spectrum. Polymers are composed of light elements, making incoherent scattering strong. Consequently, the XRD spectra of polymers often contain a relatively strong noise component.

When x-ray beams with the wavelength λ are incident on a crystal at an angle θ, coherent scattering from electrons in the crystal's atoms undergoes mutual reinforcement to form diffracted beams, as shown in Figure 8.10.

The path difference between the beams scattered from two neighboring planes is:

$$AP + PC = 2d \sin \theta \tag{8.13}$$

Here d is the interplanar spacing. Diffraction occurs when the path difference is equal to an integer number n of wavelengths.

$$2d \sin \theta = n\lambda \tag{8.14}$$

Here, n is called the order of diffraction. Equation 8.14 is known as the Bragg law, which defines the essential condition for x-ray diffraction. Therefore, for some fixed values of λ and d, only x-ray beams with incident angles that

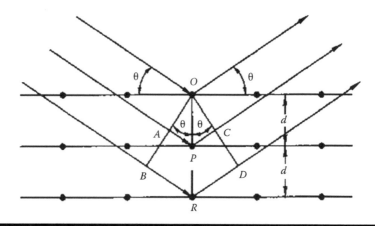

Figure 8.10 Diffraction of x-rays by a crystal.

satisfy Equation 8.15 will be reflected from a crystal. The condition on the incident angle is:

$$\sin \theta = n\lambda \,/\, 2d \qquad (8.15)$$

The direction of the diffracted beam is related to the size and shape of a unit cell, as well as to the wavelength of the incident x-ray beam. The intensity of the diffracted beam is related to atoms and their positions in a crystal. Therefore, the XRD spectrum can be considered to be a fingerprint of a crystal. This is the basis of qualitative XRD analysis.

8.3 Instrument and Specimen

A typical unit cell is on the order of one angstrom; thus, the corresponding scattering angle 2θ is in the $3–160°$ range, according to the Bragg law. Such an instrument is called a wide-angle x-ray diffractometer (WAXRD). For large crystal structures such as microcrystallines, lamellae, and spherulites, the size is on the order of 1–100 nm. For these structures, the scattering angle with respect to x-rays is quite small; thus, small-angle x-ray diffractometry (SAXRD), with 2θ in the $0.5–5°$ range, is used. In this chapter, we only discuss WAXRD.

8.3.1 WAXRD

Based on the principle of XRD, when a monochromatic x-ray is incident on a sample, the intensity of the diffracted beam can be determined as a function of angle. Therefore, a WAXRD system is composed of an x-ray tube (the light source), a goniometer (a stage with a sample on it), a radiation detector, and a record and control unit.

X-ray Tube
In Section 8.2.1 it was mentioned that x-rays are generated whenever high-speed electrons hit a metal target. Because most of the kinetic energy of these high-speed electrons is converted into heat in the target, an x-ray tube is less than 1% efficient in producing x-rays, and the metal target is required to have good thermal conductivity, and has to be constantly water-cooled to avoid melting. The intensity of the generated x-rays depends on the x-ray tube's voltage and on the atomic number of the metal target; thus, Cu has been the target of choice in many studies. The x-ray tube is covered by a heavy metal shell to prevent possible injury to the system's operator.

A filter is installed after the x-ray tube to obtain monochromatic x-rays. The atomic number of the filter material is often smaller by one than that of the target metal. Table 8.4 lists the characteristic wavelengths and absorption edges of common target metals. For example, the K absorption edge of Ni is 0.14869 nm, lying between the characteristic K_α (0.15418 nm) and K_β (0.13922 nm) lines of Cu; consequently, Ni is often used as a filter alongside Cu. Similarly, Zr can be used as a filter alongside Mo.

Table 8.4 Characteristic Wavelengths and Absorption Edges of Common Target Metals

Atomic Number	Element	$\lambda_{K\alpha}$ (nm)	$\lambda_{K\beta}$ (nm)	K Absorption Edge λ_K (nm)
24	Cr	0.22909	0.20848	0.20702
25	Mn	0.21031	0.19102	0.18964
26	Fe	0.19373	0.17565	0.17429
27	Co	0.17902	0.16208	0.16072
28	Ni	0.16591	0.15001	0.14869
29	Cu	0.15418	0.13922	0.13806
42	Mo	0.07107	0.06323	0.06198

The x-rays that are used in XRD are easily absorbed by the human body and thus are harmful, but their negative effect can be reduced significantly by properly designing the experimental equipment and user safety measures.

Goniometer

For specific locations of the line x-ray source S, the aperture and the specimen, the focusing point G of the x-rays diffracted by the specimen, S, and the specimen AB, can be located on the same focusing circle. The focusing circle is located on a diffractometer circle, with the specimen at the center. This arrangement allows for synchronous rotations of the specimen and detector. When the specimen rotates by an angle θ, the detector rotates by an angle 2θ, which ensures that the diffracted x-rays are focused on the detector, as shown in Figure 8.11.

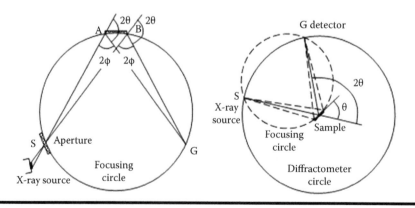

Figure 8.11 Schematic of the goniometer for determining the diffracted intensity at the Bragg angle θ.

X-ray Detector

The x-ray detector converts the incoming x-rays into electric signals. The intensity of these signals is proportional to the intensity of x-rays. Proportional counters, scintillation counters, and semiconductor detectors are often used as detectors. State of the art semiconductor detectors have the best resolution [2].

The proportional counter consists of a cylinder filled with an inert gas. The incoming x-rays are absorbed by the gas, accompanied by the ejection of photoelectrons and recoil electrons. Consequently, the gas is ionized; the so-formed electrons move toward the anode under the influence of the electric field, and positive gas ions move toward the cathode. Thus, the pulse of current reports the intensity of x-rays.

The scintillation counter exploits the ability of x-rays to yield fluorescence from fluorescent substances. The amount of light emitted is proportional to the intensity of x-rays and can be measured using a photomultiplier.

The semiconductor detector consists of two conducting electrodes and a semiconductor crystal between them. The incoming x-rays are absorbed and generate free charge carriers, i.e., electron–hole pairs, in the semiconductor. The number of carriers is proportional to the energy of absorbed x-rays. The drift of the generated electrons toward the electrodes under the action of the electric field yields pulses of current.

8.3.2 Specimen

Polymer bulk specimens are required to have a flat and smooth surface, with the thickness of 2.4–3.0 mm or higher. The diameter of fiber specimens is usually very small; consequently, a bundle contains many parallel filaments. Film and sheet specimens are required to be at least 0.5 mm thick (0.8–1.0 mm is even better); otherwise the intensity of diffracted x-rays will be too weak. Crystal powders are often filled in a sample frame, and the surface is smoothed. The diameter of powder particles is typically 10^0–10^1 μm (1–5 μm is better).

8.4 Applications in Polymer Materials

8.4.1 Crystal Form and Crystallinity

XRD is a convenient and powerful method for analyzing crystallized polymers. This method also allows to easily distinguish between different crystal forms of a polymer. For example, PP can easily form monoclinic α-crystals and hexagonal β-crystals. α-PP is thermally stable, with a high modulus but poor toughness. β-PP has a high heat distortion temperature and good toughness. Their XRD spectra are obviously different, as shown in Figure 8.12. The characteristic peaks of α-PP appear at $2\theta = 14.2°, 17.1°, 18.6°, 21.2°,$ and $21.9°$, corresponding to the (110), (040), (130), (111), ($\bar{1}$31), and (041) planes. The characteristic peaks of β-PP appear at $2\theta = 16.1°$ and $21.2°$, corresponding to the (300) and (130) planes.

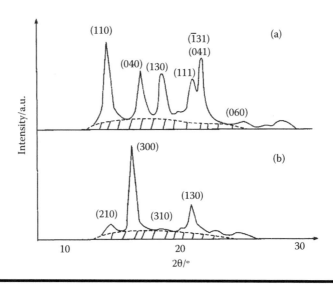

Figure 8.12 XRD spectra of (a) α-PP and (b) β-PP. The dashed areas represent the amorphous phase. (From Mo, Z., and H. Zhang. *Structure of Crystalline Polymers by X-ray Diffraction.* **Science Press, 2003.**)

Many inorganic fillers and additives used in polymer materials are crystals, and thus can be analyzed using XRD.

XRD is often utilized to determine the crystallinity X_c of polymers:

$$X_c = \frac{I_c}{I_c + KI_a} \times 100\% \tag{8.16}$$

Here, I_c and I_a are integral diffraction intensities of crystal fraction and amorphous fraction, respectively. K is the calibration coefficient. Therefore, correctly resolving the diffraction curve into crystal and amorphous phase fractions is a key problem.

According to Equation 8.16, for a homopolymer, the crystallinity equation can be rewritten as

$$X_{c,x} = \frac{\sum_i C_{i,hkl}(\theta) I_{i,hkl}(\theta)}{\sum_i C_{i,hkl}(\theta) I_{i,hkl}(\theta) + \sum_j C_j(\theta) I_j(\theta) k_i} \tag{8.17}$$

Here, i and j are the numbers of crystal and amorphous peaks; $C_{i,hkl}(\theta)$ and $I_{i,hkl}(\theta)$ are the calibration factors and the integral diffraction intensities of *hkl* planes; $C_j(\theta)$ and $I_j(\theta)$ are the calibration factors and the integral scattering intensities of amorphous peaks. Calibration factors and crystallinity equation of typical polymers can be found in the literature [3]. For example, the crystallinity equation of PE is:

$$X_c = \frac{I_{110} + 1.42 I_{200}}{I_{110} + 1.42 I_{200} + 0.65 I_a} \times 100\% \tag{8.18}$$

8.4.2 Interplanar/Interlayer Spacing

Various lamellar fillers, e.g., montmorillonite (MMT), are widely used in polymer materials. MMT is a layered silicate, with the layer thickness of ~1 nm and interlayer spacing of ~1 nm. Before it is introduced into polymer materials, organic treatment should be performed to enlarge the interlayer spacing and intercalate organics or polymer chains into it, or even to exfoliate the layers in order to obtain nanoplatelets. By dispersing the intercalated or exfoliated MMT particles in a polymer matrix, a nanocomposite is prepared. Whether MMT is intercalated or exfoliated, and the interlayer spacing of MMT can be determined using XRD.

According to the Bragg law, the interlayer spacing of MMT for the first order diffraction is:

$$d = \frac{\lambda}{2 \sin \theta} \tag{8.19}$$

Figure 8.13 shows the XRD spectra of MMT, organic treated MMT (OMMT), and its nanocomposite. The interlayer spacing is 1.23 nm. After treatment, the

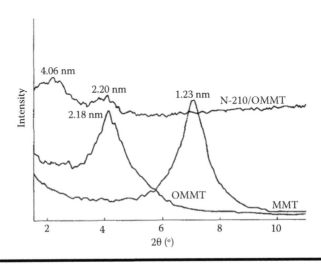

Figure 8.13   XRD spectra of MMT, OMMT, and its nanocomposite.

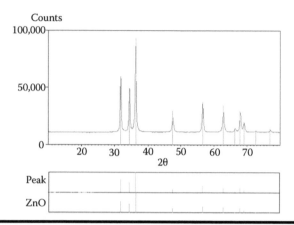

Figure 8.14   XRD spectrum of ZnO and the result of the database search.

interlayer spacing increases to 2.18 nm. After blending with the polymer matrix, most OMMT particles are exfoliated, and exfoliated nanoplatelets exhibit no diffraction peaks. However, there still are some intercalated particles, with the interlayer spacing of 2.20 nm and 4.06 nm, respectively.

8.4.3 Crystal Structure

In Section 8.1, the relations between the interplanar spacing, plane index, and lattice parameters were introduced for various crystal systems. If the crystal form of a polymer is known to belong to a specific crystal system, the corresponding lattice parameters can be calculated from its interplanar spacing.

In addition, database querying of diffraction angles, and thus interplanar spacing of strong peaks in an XRD spectrum, is a more convenient way to determine the crystal structure. Figure 8.14 shows the XRD spectrum of ZnO and the result of the database search.

References

1. Cullity, B.D. *Elements of X-ray Diffraction*, 2nd ed. Reading, Massachusetts: Addison-Wesley Publishing Company, Inc., 1978.
2. Kasai, N., and M. Kakudo. *X-ray diffraction by Macromolecules.* Berlin, Heidelberg: Kodansha Ltd. & Springer-Verlag, 2005.
3. Mo, Zhishen, and Hongfang Zhang. *Structure of Crystalline Polymers by X-ray Diffraction.* Beijing, China: Science Press, 2003.

Exercises

1. What is the origin of the short-wavelength limit of x-rays generated by an x-ray tube? Calculate the short-wavelength limit for the tube voltage of 30 kV.
2. The wavelength of the Mo K_α line is 0.071 nm. What is the energy (in eV)?
3. Calculate the plane indices of the red planes in the following pictures:

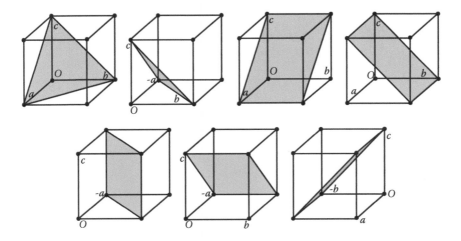

4. Determine the interplanar spacing range of crystalline polymers when wide-angle XRD is used (only first order diffraction to the Cu K_α line is considered). Is wide-angle XRD suitable for studying polymer spherulites (typically 10^0–10^1 microns in diameter)? Why?
5. Calculate the mass absorption coefficients of PVDF, PMMA and PS for the Cu K_α line.
6. According to the absorption behavior of materials with respect to x-rays, suggest methods to protect the human body from x-ray damage.

Appendix I: Names of Analytical Methods and Abbreviations

Analytical Method	Abbreviation
Atomic force microscopy	AFM
Auger electron spectroscopy	AES
Crossed polarization microscopy	CPM
Differential scanning calorimetry	DSC
Fourier transform infrared spectroscopy	FTIR
Gas chromatography	GC
Gel permeation chromatography	GPC
Inverse gas chromatography	IGC
Liquid chromatography	LC
Mass spectrometry	MS
Nuclear magnetic resonance spectrometry	NMR
Phase contrast microscopy	PCM
Pyrolysis gas chromatography	PGC
Raman spectroscopy	Raman
Scanning electron microscopy	SEM
Size-exclusion chromatography	SEC

(Continued)

Analytical Method	Abbreviation
Transmission electron microscopy	TEM
Ultraviolet-visible spectroscopy	UV-Vis
X-ray diffractometry	XRD
X-ray fluorescence	XRF
X-ray photoelectron spectroscopy	XPS

Appendix II: Names of Polymers and Abbreviations

Polymer	Abbreviation
Acrylonitrile-butadiene-styrene copolymer	ABS
Ethylene–propylene copolymer	EP
Ethylene-propylene-diene copolymer	EPDM
Ethylene–vinyl acetate copolymer	EVA
Ethylene-vinyl alcohol copolymer	EVOH
High-density polyethylene	HDPE
Linear low-density polyethylene	LLDPE
Low-density polyethylene	LDPE
Methyl acrylate–styrene copolymer	PMA-S
Methyl methacrylate–butadiene–styrene terpolymer	MBS
Natural rubber	NR
Nitrile rubber	NBR
Polyacrylic acid	PAA
Polyacrylonitrile	PAN
Polybutadiene	PB
Polybutadiene rubber	BR
Polybutyl methacrylate	PBMA

(Continued)

Polymer	Abbreviation
Polycarbonate	PC
Polyethyl acrylate	PEA
Polyethylene	PE
Polyethylene glycol	PEG
Polyethylene oxide	PEO
Polyethylene terephthalate	PET
Polyimide	PI
Polyisoprene	PIP
Polyisopropyl acrylate	PiPA
Polymethyl acrylate	PMA
Polymethyl acrylonitrile	PMAN
Polyoxymethylene	POM
Polyphenyl ether	PPE
Polypropylene	PP
Polystyrene	PS
Polytetrafluoroethylene	PTFE
Polytrifluorochloroethylene	PTFCE
Polyvinyl acetate	PVAc
Polyvinyl alcohol	PVA
Polyvinyl chloride	PVC
Polyvinylidene chloride	PVDC
Polyvinylidene fluoride	PVDF
Polyvinylpyrrolidone	PVP
Poly(methylmethacrylate)	PMMA
Poly(p-phenylene benzobisthiazole)	PBO
Poly(vinylidenefluoride-co-hexafluoropropylene)	P(VDF-HFP)
Styrene-p-fluorostyrene copolymer	PFS

Appendix III: Infrared Spectra of Typical Polymers and Characteristic Peaks

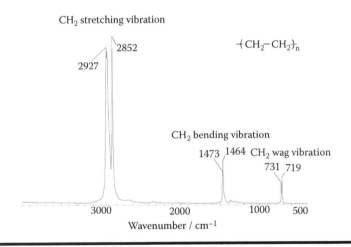

Figure a-01   IR spectrum of polyethylene (PE).

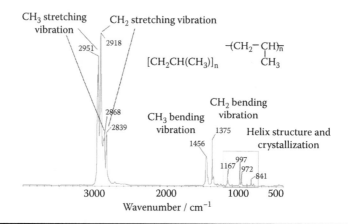

Figure a-02   IR spectrum of polypropylene (PP).

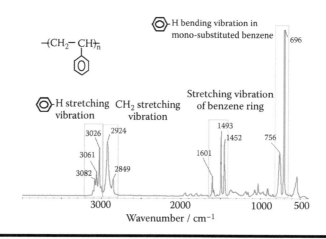

Figure a-03   IR spectrum of polystyrene (PS).

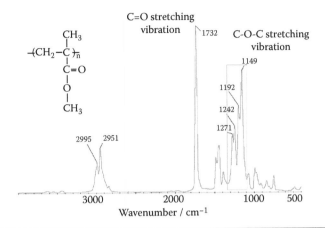

Figure a-04 **IR spectrum of polymethyl methacrylate (PMMA).**

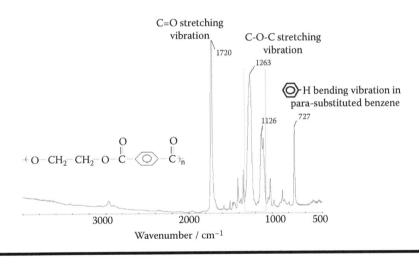

Figure a-05 **IR spectrum of polyethylene terephthalate (PET).**

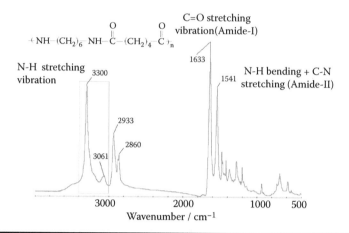

Figure a-06   IR spectrum of polyhexamethylene adipamide, nylon-66 (PA66).

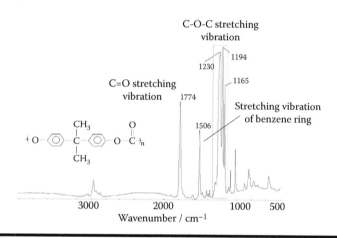

Figure a-07   IR spectrum of polycarbonate (PC).

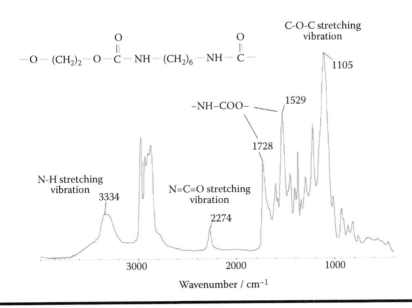

Figure a-08 **IR spectrum of polyurethane (PU) with remained isocyanate monomer.**

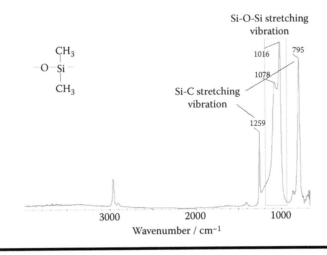

Figure a-09 **IR spectrum of polydimethyl siloxane (PDMS).**

Index

T - #0120 - 111024 - C318 - 234/156/15 - PB - 9780367572358 - Gloss Lamination